# De Havilland DH89
# Dragon Rapide

## Graham M. Simons

LONDON
IAN ALLAN LTD

First published 1986

ISBN 0 7110 1586 4



Published by Ian Allan Ltd, Shepperton, Surrey; and printed by Ian Allan Printing Ltd at its works at Coombelands in Runnymede, England

*Front cover, top:*
**Michael Astor's airworthy G-AHGD *Women of the Empire*.** *Peter R. March*

*Front cover, bottom:*
**Former British European Airways Corporation G-AGSH seen at St Just.** *Peter R. March*

# Contents

**Thanks**

In the preparation of this book many people, groups and organisations have provided much help and advice. It would be impossible to thank them all individually; however, I would especially like to thank the following:

All DH89 Dragon Rapide owners, past and present, especially Mike Kimbrell, Monique Agazarian, Bill Pinney, George LeMay, Harry van der Meer, Air New Zealand, British Airways, Air France and Finnair. The Aviation Authorities of Great Britain, South Africa, Italy, Fiji, Romania and Canada. Various aviation museums around the world, including the Musée d'Or, Portugal; Aviodome, Holland; USAF Museum, Dayton, Ohio, USA; Musée de Aeronautica y Astronautica, Madrid; South African Air Force Museum; Royal Air Force Museum, Hendon; Army Air Corps Museum; Swiss Transport Museum, Lucerne; Smithsonian Institution; Musée de l'Air, Brussels; Technichs Museum, Vienna; Canada's National Aviation Museum; and Western Aviation Museum.

The Aviation Historical Societies of Sweden, New Zealand, Australia, Finland, Belgium and the Netherlands. Bill Byard; Dave Filby; Bill Oliver; Lettice Curtis; Diana Banarto Walker; Eric Farrington Birchall; Air Vice Marshal J. de M. Severne, Captain of the Queen's Flight, Royal Air Force; Stuart Howe, de Havilland Aircraft Museum; Vernon Clarkson, British Aerospace PLC; all members of the East Anglian Aviation Society, especially Rod Beeton, Roger Wastell and the late Alan Burfield.

Finally special thanks to my wife Anne, without whose help and patience none of this would have been possible.

*Graham Simons*

**Simple maintenance requirements and the ability to operate from small strips ensured the success of the de Havilland DH84 Dragon.**
*British Aerospace*

# Evolution

The concept of the de Havilland DH89 came about as the result of three previous designs, all of which were intended to provide commercial operators with an economic aircraft capable of paying its way in an era of what was by today's standards comparatively light traffic. The first design to really fulfil this need was the DH83 Fox Moth which first appeared in 1932. Seating was for four passengers plus a pilot. The Fox Moth was powered by the four-cylinder inverted air-cooled DH Gipsy III engine, producing 120hp; it used the earlier DH82 Tiger Moth mainplanes and other components, and cost £1,045. Even when fitted with the 130hp Gipsy Major the type produced a remarkable 4½ capacity ton miles per gallon of fuel at a sedate 105mph.

The outstanding success of the Fox Moth made one operator, Mr Edward Hillman, owner of Hillman Airways, urge de Havilland to consider the idea of a twin-engined version, suitable for carrying six passengers and their luggage on Hillman's London to Paris service. A request for a machine matching this description had already been made to de Havilland's by the Iraqi Air Force, so when Mr Hillman was shown A. E. Hagg's design for the new aircraft he ordered four machines direct from the drawing board.

Thus the DH84 Dragon came into being. The Dragon made its first flight on 24 November 1932, when G-ACAN took to the air with the Class B marks of 'E-9'; at the controls was Capt Hugh Broad, de Havilland's test pilot.

The Dragon was offered for sale at a price of £2,745 ex-works, without furnishings. The DH84 produced a figure of 5.2 capacity tons per gallon of fuel, powered by two DH Gipsy Major I powerplants mounted forward of the lower mainplanes. These gave the Dragon a cruising speed of 109mph, a range of 460 miles and a ceiling of 12,500ft.

On 28 September 1933 the Australian Government issued a specification for an aircraft design that would be capable of flying the Singapore-Darwin-Cootamunda route of the England to Australia Air Mail Service; about the same time de Havilland submitted a proposal for a four-engined airliner with high performance and a wide margin of safety in the event of an engine failure. It was soon noticed that this new design not only met the requirements laid down in the Australian specification, but would exceed the performance handsomely! A contract was placed by Imperial Airways on behalf of Qantas Empire Airways for one of the new machines, to be used as the basis for the tender to the Australian Government. QEA was formed as a subsidiary to the Queensland &

**Construction Details of DH89A**
The illustration shows the structural arrangement of stub-plane, including fuel and oil tanks, struts, undercarriage and engine mounts.

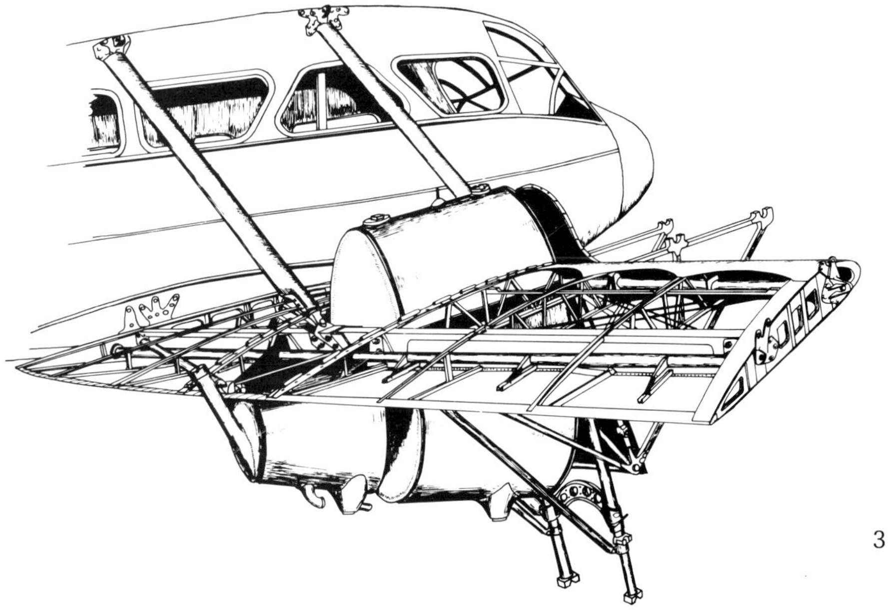

## Typical Prewar DH89 Cockpit Layout

**1** Fuel Cock Control (port); **2** Fuel Cock Control (starboard); **3** Heater Control; **4** Navigation Light Switch; **5** Mixture Control Lever (port); **6** Mixture Control Lever (starboard); **7** Throttle Lever (port); **8** Throttle Lever (starboard); **9** Tail Trim Handwheel; **10** Brake Lever; **11** Rudder Pedals; **12** Control Column; **13** Compass; **14** Fuel Contents Gauge; **15** Button Press for Fuel Contents; **16** Fuel Tank Indicator Selector Switch; **17** Altimeter; **18** Air Speed Indicator; **19** Cabin Light Switches; **20** Turn and Bank Indicator; **21** Directional Gyro; **22** Pitot Head Heater Switch; **23** Rate of Climb Indicator; **24** Suction Gauge; **25** Artificial Horizon; **26** Visual Indicator, Type I; **27** Engine RPM Gauges (port and starboard); **28** Rudder Bias Control; **29** Clock; **30** Oil Pressure Gauge (port and starboard); **31** Fire Axe; **32** Identification Switch; **33** Flap Lever; **34** Course and Height Indicator; **35** Speed Restriction Panel; **36** Fuse Panel; **37** Drift Sight; **38** Fire Extinguisher; **39** Label for Ki-Gas pump; **40** Changeover Cock for Ki-Gas Pump; **41** Ki-Gas Pump; **42** Spotlight Dimmer Switch; **43** Engine Switches; **44** Compass Light Dimmer Switch; **45** Remote Radio Controls; **46** Aerial Winch; **47** Control Cabin Light.

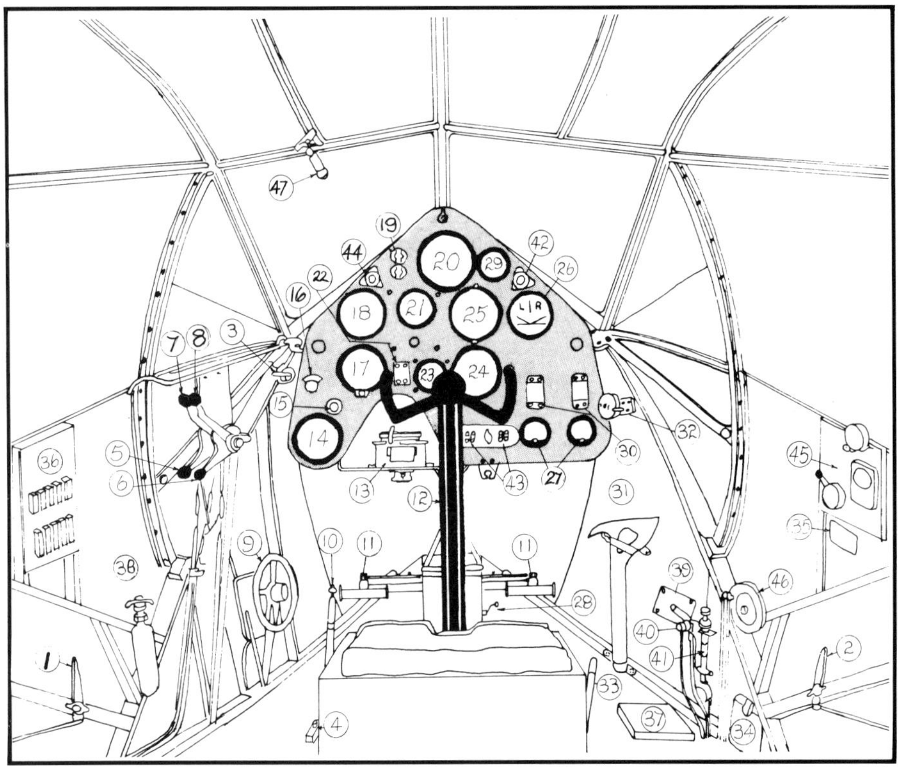

Northern Territory Air Service and was to be amalgamated with Imperial Airways for this service. A major condition of the specification was that all aircraft submitted had to possess a valid Certificate of Airworthiness (CoA) by the closing date for entrants on 31 January 1934. This proved to be well within the factory's capability for a CoA was issued to the prototype 'E-2' (later G-ACPL) and the type's first flight was made on 14 January 1934 from Stag Lane.

The type was given the designation of DH86, and was powered by a new and more powerful engine developed from the proven Gipsy Major. This six-cylinder inverted engine was given the official name of Gipsy Six, later changed by the Royal Air Force to Gipsy Queen.

Accommodation in the DH86 was offered with five different interior layouts, ranging from an eight-seat 'Long Distance Luxury Air Liner' to the 16-seat 'Air Ferry' model. An eight-berth 'Sleeping Car' model was also offered. Most were equipped as 10-seaters, although at least one is thought to have been fitted out with 18 seats. Although never given an official name by de Havilland, the type was commonly known as the Dragon Express or the Express Air Liner.

From this DH86 design, the DH89 emerged. As with all the previous designs, it was a result of the work of de Havilland's design office, headed by Arthur E. Hagg. In appearance the DH89 resembled the larger DH86. It employed the same style of fuselage, tapered mainplanes, engine nacelles and a fixed faired undercarriage, but all

scaled down to meet the twin-engined Dragon concept. Although initially called the Dragon Six, its name was soon changed to Dragon Rapide.

Built almost entirely of wood, the Rapide fuselage was a box structure with spruce longerons and struts inside a plywood covering. It featured a main cabin 13ft 6in long, 4ft 6in high and 4ft wide, had an entrance door at the rear on the port side, and the number of cabin windows was dependent on the seating arrangement. The wings were equal span, tapering towards the tips, and tapered ailerons were fitted to all four mainplanes, inter-connected by push-pull rods running inside the outer interplane struts with the connecting arms acting as mass balances. The structure of the mainplanes was one of timber spars spindled to 'I' section for lightness, wooden girder ribs and tubular steel drag struts. All were wire braced internally. The stub-planes on to which were bolted the engine mountings had tubular steel main spars and also carried the fuel and oil tanks. This whole wing structure was braced externally by struts and wires placed between the wings, engines and fuselage top. Wire-braced, the tailplane was adjustable in the air by means of a screwjack under the front spar. The rudder was horn balanced, whilst the elevators were not.

Each engine was fitted with a 'Rotax' starter, but could be swung by hand if necessary. Each engine nacelle housed a 38gal fuel tank, with a 3½gal oil tank fitted behind it; the whole nacelle was neatly faired with sheet Elektron, continuing down the undercarriage legs to form 'trousers'. Elektron is a magnesium alloy, chosen for its lightness and strength. Erroneous opinions still prevail regarding the flammability of magnesium with the attendant fire hazard. In fact solid magnesium cannot be ignited below 600°C, and even with intense localised heat the high thermal conductivity will ensure rapid cooling to help prevent the fire from spreading. Steering was differentially controlled with the rudder through a tracking tailwheel.

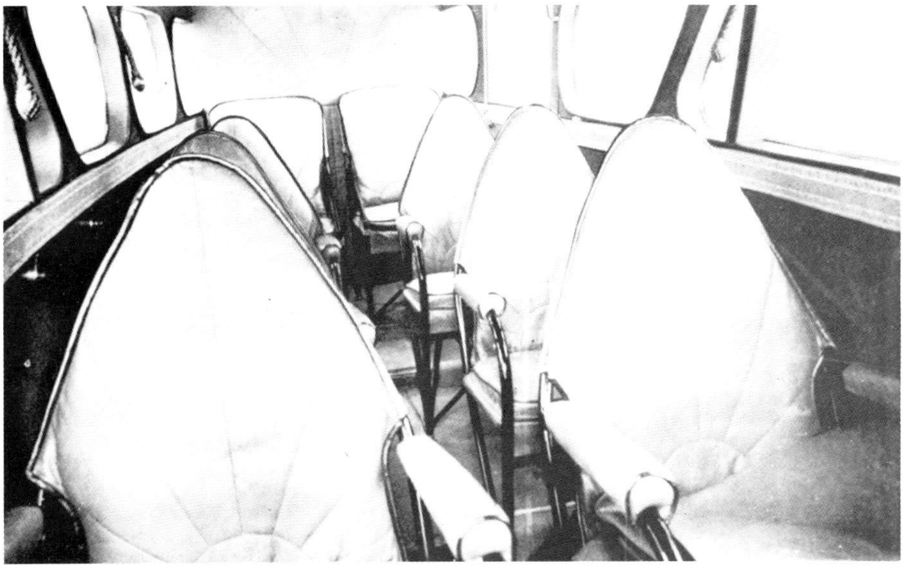

*Left:*
**The DH86 prototype powered by four 205hp Gipsy Six engines.** *British Aerospace*

*Top:*
**A standard de Havilland Dragon Rapide interior.** *R. Beeton Collection*

*Above:*
**Luxury interiors could also be specified for the Rapide; this one may well be that fitted in Olley Air Services' G-ACYR.** *R. Beeton Collection*

Powered by two 200hp Gipsy Six engines, the prototype made its first flight from de Havilland's new factory at Hatfield on 17 April 1934 in the hands of Capt Broad, carrying the Class B marks of 'E-4'.

Even while the prototype was being built, plans were in hand for series production, the first order coming from Hillman Airways. The prototype's CoA trials took place at Martlesham Heath during May and June 1934 at an all-up-weight of 5,000lb with seating for six passengers and 140lb of luggage. The report of these trials stated that the two Gipsy Sixes fitted, Nos 6008 and 6009, were limited to a maximum of 2,350rpm. Performance trials discovered that the aircraft took 1min to reach 1,030ft, 2min to reach 2,000ft and 3min to reach 2,950ft. During the take-off the aircraft attained a height of over 200ft after travelling a distance of 656yd from rest. The take-off run was 300yd, with a landing run of 500yd without brakes, 265yd with.

## Stub-Plane and Lower Mainplane of DH89A Dragon Rapide

**1** Rib 1; **2** Rib 2; **3** Rib 3; **4** Rib 4; **5** Rib 5; **6** Rib 6; **7** Rib 7 (see note); **8** Rib 8 (see note); **9** Rib 9; **10** Rib 10; **11** Rib 11; **12** Rib 12; **13** Bracing Tube; **14** Flap Torque Tube; **15** Timber Rear Spar throughout; **16** Rib 16; **17** Rib 17; **18** Rib 18; **19** Outer Strut Attachment Point; **20** Outer Drag Strut; **21** Plywood Leading Edge Covering, with local strengthening around Outer Strut Attachment and Wingtip; **22** Drag Strut; **23** Anti-Rubbing Pads laced between each Bracing Wire; **24** Timber Main Spar; **25** Internal Bracing Wires; **26** Metal Structure; **27** Steel Main Spar in Stub-Plane.

*Note:* According to the de Havilland maintenance manual there are two Rib 7s and two Rib 8s. The Ribs 7 and 8 of the stub-planes are *not* the same as Ribs 7 and 8 of the extension plane because they have different part numbers. In addition, if there ribs were the same, the wing would have a 'stepped' section at this point — which is not possible because each and every rib is slightly shorter in length than its inboard neighbour.

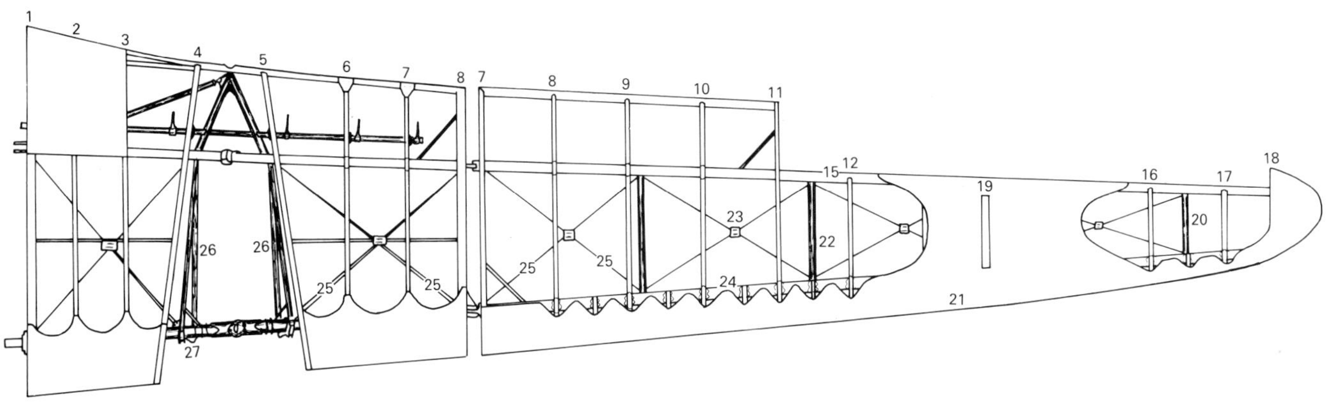

## Upper Mainplane of DH89A Dragon Rapide

**1** Rib 1; **2** Rib 2; **3** Rib 3; **4** Rib 4; **5** Rib 5; **6** Rib 6; **7** Rib 7; **8** Rib 8; **9** Rib 9; **10** Rib 10; **11** Rib 11; **12** Rib 12; **13** Bracing Tube; **14** Drag Strut; **15** Outer Strut Attachment Point; **16** Rib 16; **17** Rib 17; **18** Rib 18; **19** Outer Drag Strut; **20** Plywood Leading Edge Covering with local strengthening around Outer Strut Attachment and Wingtip; **21** Anti-rubbing Pads laced between each Bracing Wire; **22** Spar Joints; **23** Internal Bracing Wires (two each bay); **24** Timber Main Spar; **25** Cut-out in Leading Edge for Wind-driven Generator; **26** Bracing Struts; **27** Timber Rear-spar.

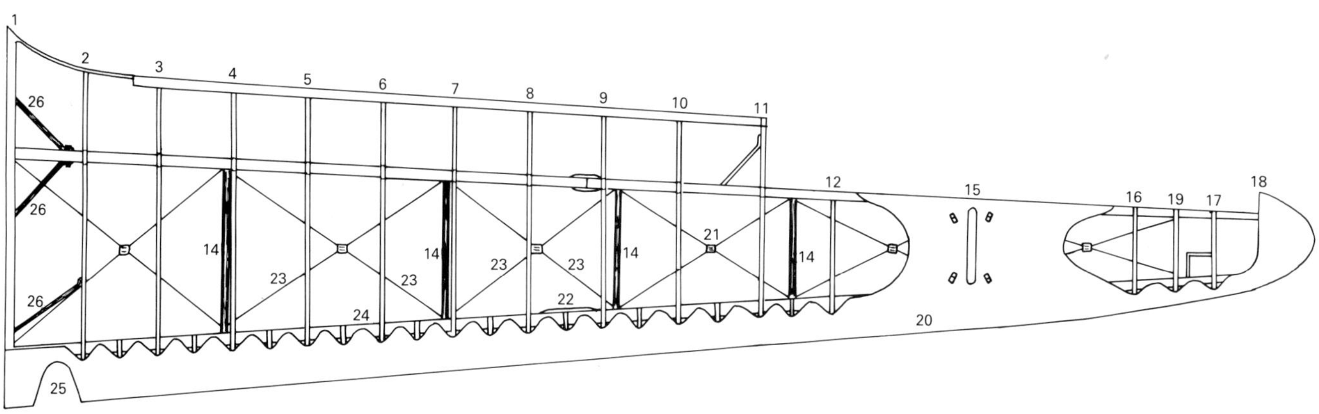

At 6,000ft at 82-86mph IAS the aircraft was found to fly level with either engine off. The controls were light and effective, with the rudder and ailerons becoming slightly heavy as speed increased.

Handling tests covered the entire speed range, with both engines on and off, the centre of gravity in both the foremost and aftermost positions. Turning manoeuvres were normal at all speeds, engines on and off, including turning against either engine at full throttle with the opposite engine off. Recovery from side slip was normal. It was found that the aircraft could not be flown straight without a load on the rudder with either engine off and it was

*Below:*
**The prototype Rapide, or Dragon Six as it was named at first, eventually saw service with Swissair, registered as HB-APA. It is seen here in the airline's colours, flying over the Alps.** *R. Beeton Collection*

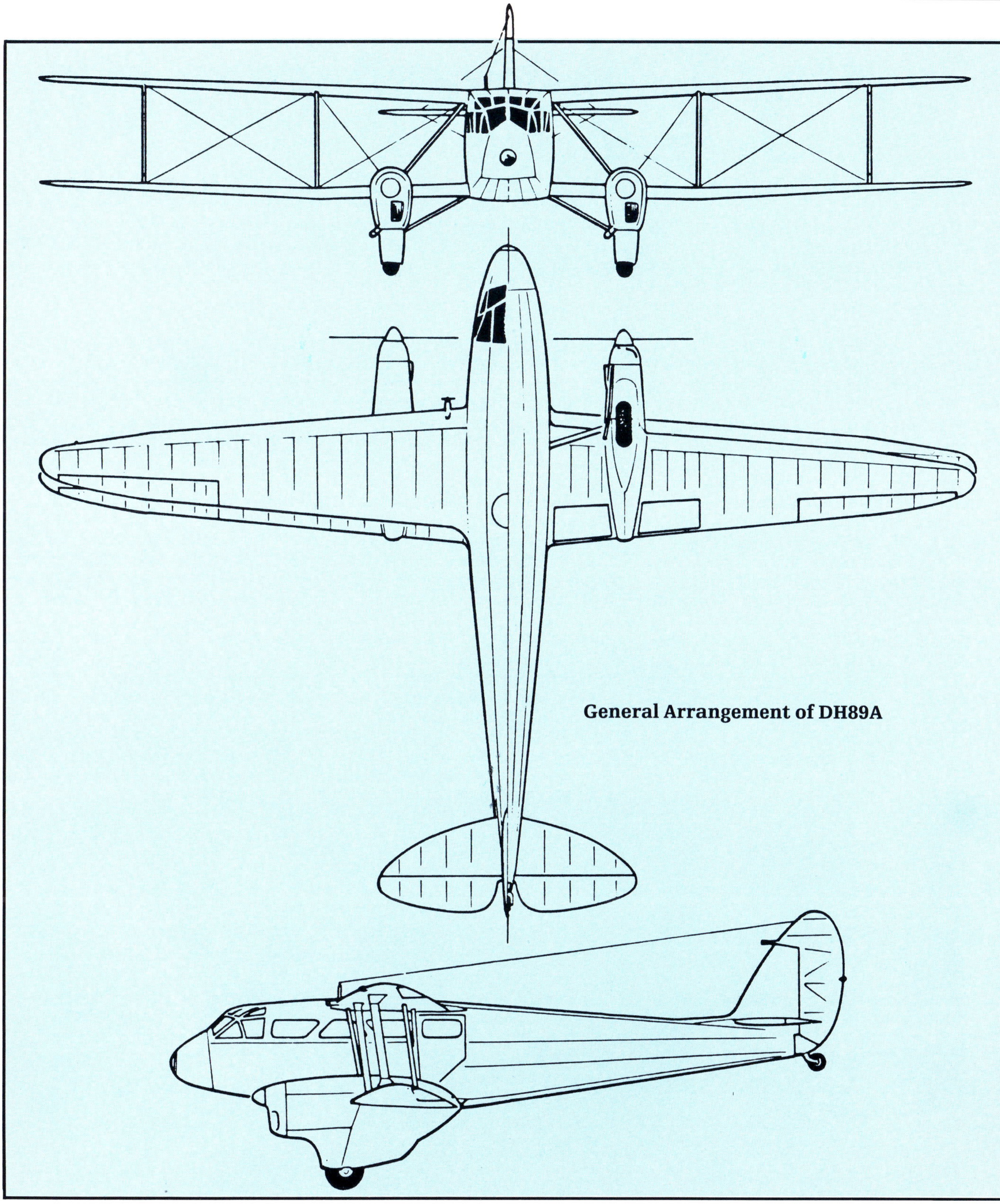

General Arrangement of DH89A

also noted that the machine could not be flown trimmed with engines on and the centre of gravity at the aftermost position.

Take-offs and landings were considered normal, with no tendency to swing or drop a wing. The slowest gliding-in speed was 63mph IAS, and such a descent was found to be very flat.

Finally, the report stated that the general design and construction was good, with the exception of the wheel fairings which had insufficient ground clearance. The cockpit was thought to be very comfortable, but the entry and exit to the cockpit was very restricted from the main cabin.

Various faults were found during these tests that it was felt should be put right. The tail trimming gear was geared too low, while it was considered that the electric starter buttons should be mounted on a more rigid plate. The mixture controls were stiff to operate, and it was suggested that arrangements should be made for stopping the engines externally, as at the time it was necessary to remove part of the engine cowlings for this purpose.

After completion of these tests the aircraft was returned to Hatfield before being sold to Herr G. Herzig of Ostschweiz Aero Gesellschaft as CH-287 for use on services between St Gaul, Zurich and Berne.

# Early Operation and

The first public display of the Dragon Six was made by the first production machine G-ACPM, when it was entered into the 1934 Kings Cup Air Race by Viscount Wakefield of Hythe. The race commenced on 13 July but the aircraft, piloted by Capt Broad, had to withdraw after heat nine of round two, when the wings were severely damaged by hailstones over Waddington, Lincolnshire. Three months later tragedy struck G-ACPM when on 20 October it flew into the sea off Folkestone, due to rain, low cloud and poor visibility. All seven occupants lost their lives.

A standard Dragon Six was ordered by the New Zealand Centenary Air Race Committee and was nominated for entry into the MacRobertson England to Australia air race by Oliver Nicholson, the President of the New Zealand Committee. Registered ZK-ACO and issued the race number '60', the aircraft was named *Tainui* after the first Maori war canoe to reach New Zealand. The aircraft was fitted with three semi-cylindrical fuel tanks in the main cabin holding a total of 240gal, increasing the range by 1,000 miles.

With Sqn Ldr J. D. Hewett as the pilot along with Mr C. F. Kay, and Mr E. Stewart as the Radio Operator, ZK-ACO was positioned at Mildenhall for the start of the race on 20 October. It was to be the 11th aircraft to take-off for Melbourne.

The race was not without surprises for the crew, including a forced landing at Boulogne, and wing damage suffered at Cloncurry when Hewett and Kay taxied the aircraft into a fence. Luckily the damage was repairable, and they flew on finishing a respectable sixth in the handicap and fifth in the speed race. The flying time was 87hr 54min 36sec formula time and 106hr 51min 28sec net flying time. The total time taken to get from Mildenhall to Melbourne was 330hr 51min 10sec.

Production of the Rapide soon gained momentum: 11 aircraft were completed in 1934, including aircraft for Hillman's, Ala Littoria of Italy and the Anglo Persian Oil Co. The first for Anglo-Persian, G-ACZE, left Heston Airport on 17 December and arrived at Almaza, Cairo on Christmas Day; together with G-ACZF (which was flown out in February) it was based in Abadan in the Persian Gulf. Operated on behalf of Anglo-Persian by Airwork Ltd, they were used for quick communications between the oil shipping base and the oilfields in the surrounding desert. Anglo-Persian was re-named Anglo-Iranian Oil Co Ltd in March 1935 after Persia became Iran, and the company continued to operate the Rapide until after World War 2. The first two machines were soon joined by G-AEMM, and all three aircraft were replaced in 1938 by three DH89As, the original aircraft being flown back to Heston and sold off.

The three replacement aircraft, G-AFHY, G-AFHZ and G-AFIA, were all flown out from Heston on 10 October 1938, with Capt L. A. Lafone, the overseas manager of Airwork Ltd, in G-AFIA, finally reaching Abadan on the 19th, via Paris, Marseille, Ajaccio, Tunis, Benghazi and Cairo (where Airwork did the required 25-hour inspection), then from Cairo through Bagdad to Abadan.

The DH89s waiting there to be returned were fitted with radios that had to be transferred to the DH89As, so after a change of equipment the old aircraft headed back to the UK. Capt Lafone left Abadan on 8 November in G-ACZF, landing at Bagdad for a night stop. It flew on to Rutbah Wells H5, an

# the Royal Rapides

*Left:*
**ZK-ACO, the New Zealand MacRobertson Air Race entrant, at Singapore during the race. The aircraft had a varied career, before crashing in New South Wales, Australia in 1952.** *John Horne Collection*

*Above:*
**It is believed that G-ACTT made only one flight in Anglo-Iranian Oil Co Ltd colours, illustrated here, before being transferred to the ownership of HRH Prince Edward the Prince of Wales.** *Harry van der Meer*

oil pipeline strip, then to Lydda, before a night stop again at Cairo. The rest of the route was via Amseat, Benghazi, Tripoli, Tunis, Cannes, Marseille and Lyon, arriving back at Heston on the 24th.

The leg from Tunis to Marseille was particularly interesting. After take-off from Tunis the flight was through heavy rainstorms nearly all day. As the flight progressed the controls grew steadily stiffer. It was later realised that G-ACZF had been out in the desert for nearly three years, the hot dry atmosphere causing the timber to shrink as it dried out. This had been taken into account by the ground engineers by the simple exercise of adjusting the tension in the control cables. The moisture forced into the structure during the storms made the timber quickly regain its original dimensions, causing the controls to almost seize up!

The type also formed flying equipment for other oil based concerns in the Middle East, including the Iraq Petroleum Co and its associate concern, the Societe de Transports de Proche-Orient. Elsewhere, three were used by the Asiatic Petroleum Co for communications duties between oil sites in Netherland New Guinea.

## Royal Rapides

An aircraft that was originally ordered by Anglo-Persian Oil Co was construction number 6257, the eighth Dragon Rapide built. Allocated the registration G-ACTT, it was granted a CoA to Anglo-Persian on 20 October 1934. However, before the aircraft could be delivered, ownership was transferred to His Royal Highness Prince Edward the Prince of Wales.

There is evidence to suggest that for a considerable time HRH had been investigating the suitability of the type for his own use, for on 7 February 1935 when de Havilland issued Allotment No 40 for G-ACTT there was a very long detailed list of additional equipment and requirements that included the phrase 'fittings as on machine 6261'. This was G-ACYR of Olley Air Services.

The allotment was made in response to a letter from the Prince ordering the aircraft. Although no evidence is available, it is thought that G-ACTT was diverted from the Anglo-Persian order to fulfil the wishes of the Prince to have all the aircraft in his fleet registered with double ending registrations. The customer mentioned on the allotment was Fl Lt Fielden of Royal Air Force Hendon, who was the Prince's personal pilot. A Certificate of Airworthiness 'for Public Transport' was required, with a delivery date quoted as 'within nine weeks from 7 February 1935 to fly away'.

The interior furnishing was to be exactly similar to 6261, with the exception of the standard rear bulkhead which was to be fitted with standard separate luggage and lavatory compartments. The lavatory was to be convertible to accommodate an extra single chair, exactly the same as the single chair in the aft of the main cabin, conversion being made by removal of the door. An alternative to the individual chair aft of the rear pair was for the provision of a special chair or stool for the radio operator.

The upholstery and coverings in the cockpit and cabin interior were to be the same as for No 6261, with the exception that the cabin chairs themselves be upholstered in best red leather with grey piping. The Prince of Wales feathers were to be embossed on the back of each seat. The pilot's seat was also in red leather, with the cockpit painted grey throughout. Curtains were to be fitted to the cabin windows, with rope pulls as per 6261. A cabinet with flap table — carrying two large thermos flasks, four cups and saucers, one bottle of whiskey, four tumblers and two soda syphons — was to be placed in the space normally provided for the alternative radio position. The bulkhead behind the control cabin was to have two doors.

The 9½gal auxiliary petrol tanks were to be fitted under the chairs in the main cabin, with total oil capacity to be 4½gal. An altimeter and clock were fitted in the main cabin on an instrument panel, while the radio apparatus was fitted in alternative positions, firstly above the rear of the pilot's bulkhead, and secondly recessed into the wall separating the cabin from the luggage compartment: remote control for the radio was provided for the pilot. The wireless was provided free issue from Marconi, and fitted both with fixed and trailing aerials. The instrument board was to accommodate a free issue Sperry horizon and direction finder. In addition, a turn and slip indicator (the type of which was to be specified at a later date) and a clock with a special hand tripping device, were to be mounted on a special shock absorbing panel.

An ensign staff was to be fitted to the aircraft at the rear of the control cabin and a Harley landing light to be installed in the extreme nose of the aircraft. A step ladder was provided for attachment to the lower wing trailing edge.

The whole aircraft was to be completed to the latest specifications, with all current modifications embodied, and was to be painted as same as the Prince's DH84 Dragon. The colour scheme, like all of his aircraft, was in the red and blue colours of the Guards, a scheme that made the aircraft easily identifiable. The registration, cheat line and window frames were in silver.

The delivery note, A2797, was dated 26 April 1935, and lists the serials of all the equipment fitted. One small item that appears on the delivery note but not the original allotment was for the fitment of a small notice in the main cabin proclaiming that 'Smoking permitted when Auxiliary Petrol Tanks removed'.

How much the Prince used the aircraft is not known, but it is known that G-ACTT was flown five times on the day it was delivered, as Edward Fielden's log books show. These flights were as follows:

11.00hrs Hatfield to London
14.00hrs Hendon to Smiths Lawn (Windsor Castle)
14.25hrs Smiths Lawn to Hendon with the Prince
17.05hrs Hendon to Smiths Lawn with the Prince
17.25hrs Smiths Lawn to Hendon.

Little else is known of the usage of G-ACTT until the aircraft was offered for sale at an undisclosed price. The advertisement stated that the aircraft had been flown for 51hr 15min.

One other DH89 was used by British Royalty. G-ADDD (c/n 6283) was delivered on 6 June 1935 with the same interior fit as G-ACTT. This aircraft was to be the first machine operated by the King's Flight of the Royal Air Force, which was formed on 21 July 1936. Flt Lt Fielden was the pilot, with the appointment of Captain of the King's Flight. The remaining personnel of the Flight consisted of a clerk and two mechanics. G-ADDD remained in Royal service until sold to Western Airlines, being delivered to the company by Flt Lt Fielden at Weston-super-Mare on 8 May 1937.

*Above:*
**The interior of G-ADDD of the King's Flight, the same as specified for G-ACTT.**
*The Queen's Flight*

*Below:*
**G-ADDD of the King's Flight, RAF.** *Royal Air Force Museum*

# Military Conversion and Prewar Civilian Service

In early 1935 Air Ministry Specification G18/35 called for a new class of aircraft intended for coastal reconnaissance, and to a certain extent for offensive operations against shipping. This was contended for by de Havilland with a redesign of the Rapide as a potent military aircraft. One prototype was built to Air Ministry order, and allocated the RAF serial K4772. References have appeared over the years doubting the correct designation of this aircraft, c/n 6271, the first DH89M. This designation appears in de Havilland's archives as being correct for the c/n. However, this was not the first of the Dominies, which was the name given by the RAF during World War 2.

Structurally, K4772 was very similar to its civil counterpart. The cockpit was almost the same, apart from the Vickers Mk 3 belt-fed machine gun on the starboard side, with its breech accessible to the pilot. In the front portion of the fuselage was the bomber/observer's position, and on the port side was a navigator's table. Opposite the navigator sat the radio operator, complete with transmitter/receiver, and in the rear of the main cabin, in place of the rear escape hatch, was a de Havilland-designed gun mounting and swivel seat for the gunner.

The bomb load was two 100lb and four 20lb bombs, the former being slung under the middle of the fuselage accessible from inside the aircraft via a raised box in the floor. When released, they fell through spring-loaded traps in the aircraft's belly, the flaps closing automatically after the projectiles passed through.

The airworthiness trials brought very similar results to those for the civil prototype back in 1934. It was found that the aircraft showed a tendency to drop the starboard wing during slow landings, although the military aspect of the tests demonstrated that it was a steady bombing platform, except in bumpy weather when it was very easy to over-correct. The bomb aimer's position was found to be very cold, uncomfortable and draughty, with bombing instruments and releases only convenient for the bomb aimer. The view ahead was thought to be inadequate. The efficiency of the forward gun was very good, the aircraft being a steady gun platform, with an excellent forward view. Also the free gun position was very easy of access, but very draughty.

It was expected that the bomb cell spring-loaded doors would open in flight, but it was found not to be the case, and as far as could be ascertained the doors did not alter the trajectory of the bomb. A Mk 3 bombsight was fitted for the trials, the slot for the sight being

*Above:*
**In this view of K4772 the rear upper gun position and gunner's windshield are clearly visible.** *RAF Museum*

*Below:*
**First of the DH89Ms was c/n 6271 built to an Air Ministry order and given the RAF serial K4772. Note the underside bomb-rack/bay and the nose machine-gun.**
*RAE Farnborough*

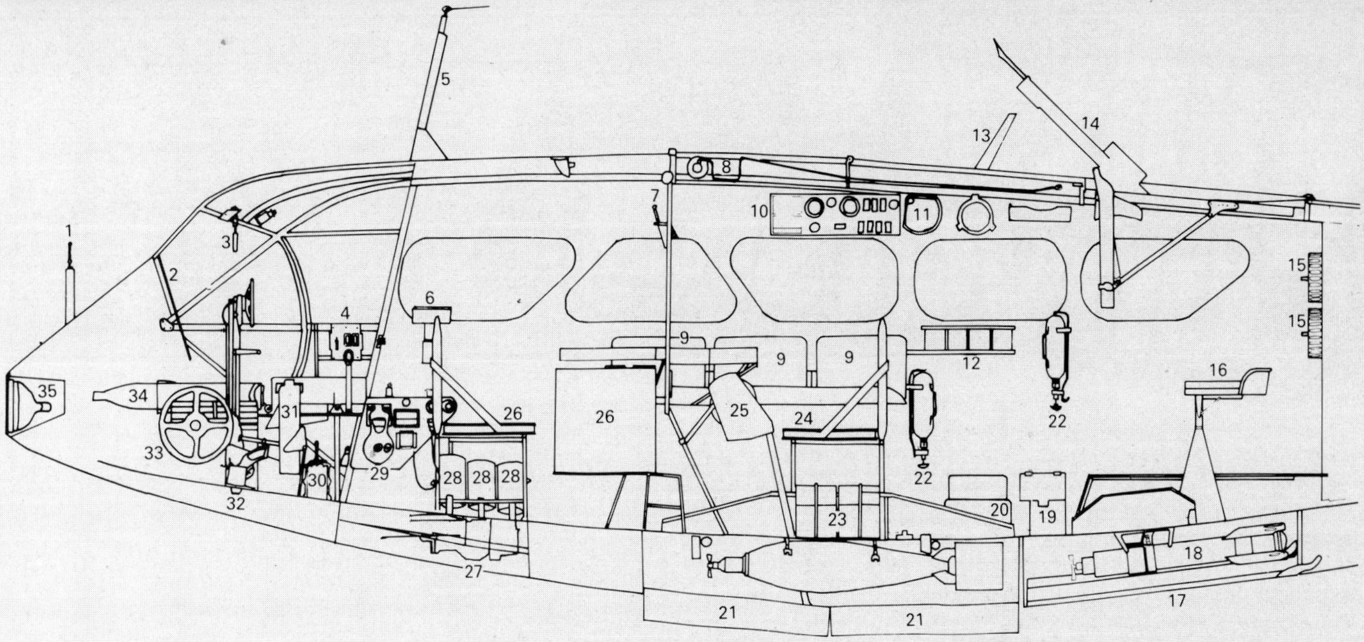

**Side View Interior of DH89M Prototype for RAF — K4772 c/n 6271**

**1** Ring and Bead Gun Sight; **2** Instrument Panel; **3** Control Column; **4** Navigation Light Switches; **5** Fixed Aerial; **6** Compass; **7** Course and Wind Calculator; **8** Stowage for Dinghy; **9** Parachute Stowage Troughs; **10** Electrical Control Panel; **11** Mounting for Aldis Lamp; **12** Stowage for Sea Markers; **13** Wind Deflector; **14** Lewis Gun and Mount; **15** Stowage of Lewis Gun Drums; **16** Top/Rear Gunner's Seat; **17** Rearwards Sliding Door; **18** Light Series Bombs; **19** First Aid Stowage; **20** Accumulator; **21** Bomb Doors in Open Position; **22** Fire Extinguishers; **23** Wave Meters; **24** W/T Operator's Seat; **25** Trailing Aerial Winch; **26** Navigator's Seat and Fold-down Chart Table; **27** Bomb Sight; **28** Flame Floats; **29** Bomb Aimer's Control Panel; **30** Bomb Fusing Lever; **31** Ammunition Box; **32** Very Pistol; **33** Tail Trim Wheel; **34** Vickers Gun; **35** Landing Lamp.

closed by a two-piece sliding panel which was found most unsatisfactory in operation, because the panel showed a marked tendency to jam, mostly when being opened.

Due to the urgency of the trials, no internal communication facilities were fitted, all instructions having to be shouted. This was obviously unsatisfactory, prompting a recommendation to fit them as soon as the preliminary trials were over.

Towards the end of 1935 K4772 was

*Above:*
**Quebec Airways operated CF-AYE on air survey work, and it is seen here in 1942, with non-standard undercarriage fairings — no trousers!** *Dave Filby*

*Right:*
**CF-AEO, with the additional main cabin door and extended dorsal fin as fitted to all Canadian DH89s.** *National Museum of Air & Space, Canada*

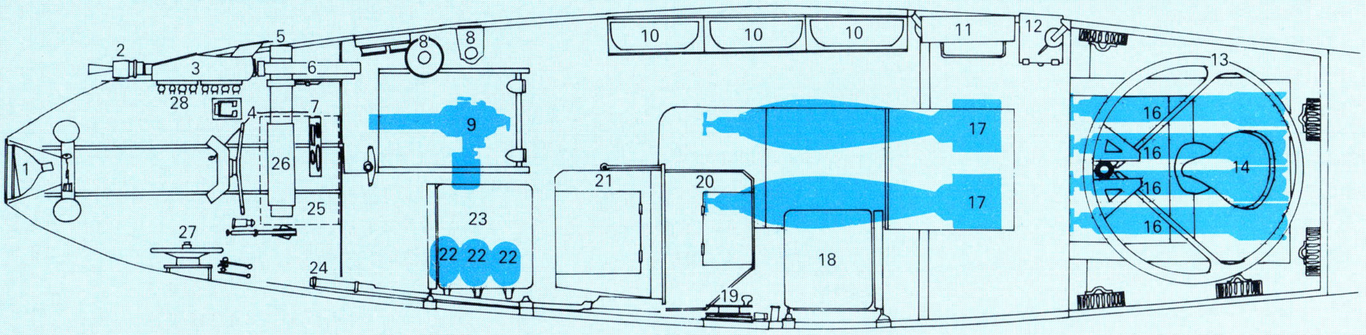


**Plan View Interior of DH89M Prototype for RAF — K4772 c/n6271**
**1** Landing Lamp; **2** Vickers Gun and Chute; **3** Cooling Tube; **4** Trigger Release; **5** Link Chute; **6** Empty Case Chute; **7** Bomb Aimer's Instrument Panel; **8** Compass and Mount; **9** Bomb Sight under Hatch (sliding panel under Bomb Sight); **10** Parachute Stowage Trough; **11** Accumulator; **12** First Aid Box; **13** Lewis Gun Ring and Mount; **14** Rear/Top Gunner's Seat; **15** Lewis Gun Ammunition Drums; **16** Light Series Bombs (under floor); **17** Bombs (under floor); **18** W/T Operator's Folding Seat; **19** Aerial Winch; **20** W/T Operator's Table; **21** Navigator's Table; **22** Flame Float Stowage; **23** Navigator's Fold-down Seat; **24** Cabin Heat Duct; **25** Pilot's Seat (shown dotted); **26** Ammunition Box; **27** Tail Trim Wheel; **28** Stowage of Very Pistol and Cartridges.

sent to the Coastal Defence Development Unit at Gosport for comparative trials with its rival, the Avro 652, later to be known as the Anson. The DH89M failed these trials, the contract going to Avro, but the work on K4772 was not wasted. From Gosport the aircraft went to the Royal Aircraft Establishment at Farnborough on 19 June 1935, and by July it had been allocated to the Instrument & Photographic Flight. By the end of September K4772 had been transferred to the Wireless & Electrical Flight for use on general radio telephony work and landing lamp tests.

By late spring of 1936 it was back with the Instrument & Photographic Flight for a series of tests on auto control research. These trials eventually led to the design and manufacture of the Mk 8 autopilot in 1942, many of the stability and gyroscopic lateral control tests being conducted in the aircraft.

The military conversion of the Rapide eventually came to fruition when in December 1935 the Spanish Government received three DH89Ms for police duties in Morocco, although they were based at Getafe Airfield, Madrid. The Spanish Air Force DH89Ms were fitted with a 7.7mm Vickers 'E' machine gun in the nose, a Vickers 'F' machine gun in the rear position, with a third gun firing through a trap in the floor. A Spanish-designed bombing window was installed locally, presumably to cure the problems found on K4772. Under the fuselage a bomb rack was fitted, carrying 12×12kg bombs in three groups of four.

*Above:*
**The Toronto *Globe & Mail* had owned CF-BBG, fitted with EDO floats, for only two months when it was destroyed by fire.** *EDO Floats*

The aircraft were ferried out from Great Britain with British registrations, but upon arrival in Spain were issued the general code for liaison and light bombing aircraft, 'LR:'. Each aircraft was also given an individual number (eg LR:1). Information regarding any aircraft connected with the Spanish Civil War has proved very difficult to trace, but it has been possible to piece together some details, mainly through the kind help of the Spanish Museum of Air & Space.

Two days before war broke out, two of these aircraft had been transferred to airfields that fell into Nationalist hands. The third was seized by the Nationalists when it landed at Saragossa on 18 July 1936, the machine having flown in from Madrid with General Nunez de Prado, Director of Aviation, who was due to see General Miguel Cabanallas, commander of the 5th Division. Nunez de Prado was arrested and, along with his ADC, later shot.

Three other DH89s had been purchased in Great Britain by Snr Juan de la Cervia, G-ACPN, G-ADAO, G-ADFY arriving in Spain before 13 August 1936. These aircraft were gathered together into the Dragon/Fokker Group, moving from Saragossa to Olmedo on 19 August. Unfortunately, one of the civilian Rapides was shot down in error by a German pilot flying a He-51 wrongly assuming it to be a Republican aircraft, on 27 August and another caught fire at Aragon on 13 September. The last of the civilian DH89s was used by the Commander of Romeo Co-operations Group, receiving the markings 20:2. It was later assigned to the Air Transport Group, coded 40:2, and was later converted at Burgos into a fighter with a fixed machine gun in the nose. It is reported by some sources as being a DH89M. In fact this machine was ex-G-ADAO, c/n 6275. It is pictured in *Air War over Spain* by Jesus Salas Larrazabal (Ian Allan, 1968), showing the aircraft named *Capitán Vela*, with a machine gun in the nose. What proves this aircraft to be No 6275 and not one of the DH89Ms is that the photograph clearly shows the standard 'Rapide' fin, not the extended vertical fin of the DH89M.

It is possible that one of the other two DH89s purchased in England by Juan de la Cervia was named *Capitán Pouso*.

Another Rapide involved in the Spanish Civil War indirectly but very importantly was Olley Air Services' G-ACYR. Hired from Olley's by Juan de la Cervia and newspaperman Luis Bolin for £1,010, G-ACYR was flown from Croydon to Las Palmas by Capt Charles Bebb, leaving at 08.00hrs on 11 July 1936. First stop was Bordeaux, where they were met by a group of Spaniards headed by the Marquis de Merito. Bebb flew them to Lisbon where there was a meeting with more Spaniards, before flying on to Casablanca and then to Las Palmas in the Canaries on 14 July.

Four days later Capt Bebb was woken at 03.00hrs and told to be ready to leave for Morocco, but was kept waiting until 17.30hrs, when General Franco appeared for the return flight to Casablanca. Very early the next morning they headed for Tetuán in Spanish Morocco.

After this flight G-ACYR returned to Great Britain, where it was used on normal routes until after World War 2, being stored dismantled at Desford. It was eventually presented to Franco by Air Couriers Ltd in 1953 for preservation in the embryonic Spanish Air Force Museum, where it is on display today.

Concluding the DH89's involvement in the Spanish Civil War were 10 machines purchased by the Republican side in late 1938/early 1939 for use on liaison duties. All were civilian machines, flown out of the United Kingdom with French civil registrations and with the first CoA in the name of Paul Legestelois.

At the end of the war 11 of the 13 Rapides used by the Republican side were numbered from 40:3 onwards, later being granted civil registrations and used by Iberia Airlines, at least five being placed on the Seville-Tetuán-Melilla route.

### Prewar Dominion Service

In 1935 the first of 16 airframes were shipped direct to Canada for de Havilland Canada for local modification and

*Above:*
**CF-BNG on floats, with beaching gear attached.** *Canadian National Archive*

re-sale. These Canadian machines were easily distinguished by the extended dorsal fin and undercarriage attachments for either wheels or skis, depending on weather conditions: the numerous lakes and rivers in Canada made the use of floats a natural optional choice during the summer months. The prototype for this conversion was CF-AEO. It was delivered with a temporary CoA to Quebec Airways in June 1935, which offered it for sale at Can$23,000 on wheels, or Can$26,000 fitted with floats. It was sold to Canadian Airways of Winnipeg, but was destroyed by fire at Moncton, New Brunswick when it 'mushed' on take-off.

Subsequent Rapides in Canada were supplied to Canadian Airways which used them on West Coast services and maritime routes, and British North American Airways in Toronto and Coote Airways of Vancouver also operated the type. Quebec Airways of Montreal used some on scheduled services in the St Lawrence river area. At least two of Quebec's DH89s, CF-AYE and CF-BFP, were equipped and used for photographic survey work. One of the company's pilots, Dave Filby, recalls that this was not the only modification.

'Almost all of the survey flying was done operating out of rough, unprepared strips. This caused problems with mud and grass collecting in the undercarriage fairings. We ended up fitting floatplane type cowlings, leaving the wheels totally uncovered. CF-AYE carried the normal fuel tankage, but CF-BFP was fitted with a large auxiliary tank in the fuselage to give it a much greater range. These photographic flights could take us up to 15,000ft, which took about 45 minutes to reach. We usually operated one trip a day, lasting between 1½ and four hours.'

Second-hand aircraft were also being used by many operators. The Toronto daily newspaper *The Globe & Mail* bought CF-BBG in June 1937, only to see it destroyed by fire in the Toronto air harbour just two months later on 21 August 1937, during re-fuelling.

Several DH89s were converted to floatplanes after suitable strengthening for the attachments. The Canadian Department of Transport made it mandatory as a safety measure that a second cabin door was added when floats were used, with walkways on both wings. After tests on the floatplane prototype it was noticed that lateral stability was affected when flying with floats and therefore an extension was added to the dorsal fin to compensate. At least two types of all-metal floats are known to have been fitted, the normal type being manufactured by EDO, although CF-AEO is known to have been fitted with Canadian Fairchild 5500s and was flown to Rockcliffe, Ottawa on 18 June 1935 for type approval by the RCAF.

The first Rapide imported directly to Australia was VH-UVS for the Aviation Department of the Shell Company of Australia, and appropriately named *Spirit of Shell.*

The Royal Australian Air Force ordered one aircraft for general transport work. Taken on charge on 2 July 1935 it was allocated the serial A3-1, and immediately put to work on photographic survey duties in the Northern Territories. A3-1 did not last long with the RAAF: Christmas 1935 saw an urgent request placed before the Department of Defence asking that the aircraft be released to Holyman Airways which was suffering a chronic shortage of aircraft following two crashes of DH86s. Registered VH-UUF and named *Memma*, the aboriginal name for white cockatoo, A3-1 went into service on the Essondon to Sydney route for Holyman on 27 January 1936.

A3-1 was soon replaced by VH-UVG from de Havilland Australia. This replacement aircraft was serialled A3-2, but such was the urgency to get the machine into service to take up the survey work started by A3-1, that it departed Western Queensland still wearing its civil registration. In many ways A3-2 lead a charmed life. On 13 April 1935 it took off from Camooweal to commence survey work, but the crew became unsure of their position and were forced to make a landing in the desert when fuel ran out. During the landing the aircraft bounced

in rough scrub, suffering a badly damaged undercarriage; but luckily the three crew were uninjured. Despite a large search being mounted for the missing aircraft, it was not located until 22 April. Two days later a ground party reached the wreck and rescued the crew. The aircraft was later salvaged by a RAAF party, although only the fuselage and fittings were trucked back to No 1 Aircraft Repair Depot at Leverton for a major re-build, the remainder being left on site.

The rebuild of A3-2 was completed the following year, during which the machine was fitted with a radio, making it one of the first so equipped in Australia. It went on to do more valuable survey work until on 21 May 1937 it left Tennant Creek bound for Tanami on the edge of the Gibson Desert. This meant flying over some of the most inhospitable terrain of the continent. By 13.45hrs the crew became uncertain of their position, and A3-2 made yet another forced landing, this time on the edge of a dried-up lake bed, coming to rest undamaged. Two days later it was spotted from the air and the crew rescued. Subsequently a land party reached the site on 30 May by truck, bringing petrol to refuel the aircraft. The Rapide was towed to a clear part of the lake bed and flown out.

A3-2's luck finally ran out on 3 February 1938 as it was approaching Laverton after a local training flight. The aircraft banked sharply and a wing-tip hit the ground, the ensuing cartwheel making the aircraft a total wreck. Although in service with the RAAF for only two years, it had logged 550 adventurous flying hours.

### Design Improvements

Design modifications on the DH89 continued, to improve operating efficiency and comfort. Much of this work was done on the 60th airframe, G-ADWZ, retained by de Havilland in 1935. The rear windows were lengthened, a landing light fitted in the nose and cabin heat installed, along with a toilet. With experience gained from the DH89M the gross weight was increased to 5,500lb with suitable local strengthening. The aircraft was also fitted with thickened wing-tips as a cure to the wing dropping problem noted on some of the earlier aircraft.

G-ADWZ went to A&AEE Martlesham Heath for Certificate of Airworthiness trials during April 1935, where it was noticed that with the extra equipment fitted the centre of gravity could, unless special care was taken during loading, fall outside the aftermost limit. It had been suggested that flaps would help the Rapide 'return to earth', and in November 1936 they were fitted to the 93rd airframe, first flown as 'E-4',

*Above left:*
**C/n 6303 received its first CoA in 1935 and became SU-ABS *Helwan* with Misr Airlines of Cairo. It is believed scrapped in 1948.** *RAF Museum*

*Below left:*
**Railway Air Services' *Star of Scotia*, G-AEBX.** *RAF Museum*

*Above:*
**These three aircraft were given the ferry markings G-ADUM (C/n 6315), G-ADUN (C/n 6316) and G-ADUO (C/n 6317) for Devlet Hava Yollari of Turkey. They were registered TC-ARI, TC-BAY and TC-CAN respectively in May 1936, but their subsequent fate is unknown.**

*Below:*
**A good example of the DH89 Rapide in prewar civilian service was Western Airways' G-ADBV.** *RAF Museum*

which went to A&AEE for flight tests in March 1937, now wearing the registration of G-AEOV. The A&AEE found the flaps effective, but difficult to operate due to the design and position of the flap operating lever.

During these trials it became evident that excessive tailplane buffet was occurring under certain conditions of flight. It was considered unsafe for the trials to continue, and the aircraft was returned to the manufacturer for modification. The buffeting was thought to have been caused by an increase in tailplane area and alterations in the elevator shape, and so guideplates were fitted to the tailplane incidence jack and the tailplane bracing wires were re-positioned further outboard to reduce the overhang of the tailplane. Tests were sucessfully completed after these modifications.

Series production of the improved model, designated DH89A, commenced with G-AERN. As many of the earlier aircraft returned for overhaul, flaps were retrofitted. Another innovation tried out was when Dowty levered suspension undercarriage was fitted experimentally to c/n 6420 during September 1938, flying as 'E-9', later becoming P1764 for the RAF.

Production continued throughout 1937 with 45 aircraft produced. Amongst these machines were aircraft for Olley Air Services, the Post & Telegraph Division of the Ministry of Commerce, Latvia, several for Rhodesia and Nyasaland Airways, Wearne Brothers of Singapore, Devlet Hava Yollari, and some for the military government of China.

# To WAR

Out of all the orders received, three were to have a lasting effect on the future of the design. Back in 1935 a standard Rapide had been added to the strength of No 24 Squadron RAF as K5070; used as a VIP transport by the Air Council, it was housed alongside G-ADDD. As a direct result of the successful service of K5070, P1764 and P1765 were ordered in 1938 to Air Ministry Contract 21/38 as communications aircraft.

The second order came from Messrs Airwork Ltd with an order for nine aircraft to be fitted out as navigation trainers for C. W. Martin's School of Air Navigation at Shoreham. This followed Air Ministry policy of the time, whereby, under an expansion scheme, civilian organisations were employed for service training under Type 'A' contract. The Shoreham school later became No 6 Civil Air Navigation School, with Airwork operating the aircraft. In September 1939 the unit moved to Staverton, where it became No 6 Air Observers School. Its aircraft were camouflaged on the upper surfaces, but painted Trainer Yellow on the remainder, and all still carried their civil registrations for some while after the outbreak of war. The School's 14 aircraft were finally taken over by the RAF under contract No 34983/39 on 15 July 1940.

The third order of importance was for two Rapide trainers, serialled P9588 and P9589 for No 2 Electrical & Wireless School at Yatesbury, Wiltshire. This was to be followed up with three more contracts for trainers during 1939.

Although these orders were to be of the greatest importance for the type for military applications, many deliveries continued right up to the outbreak of hostilities to civil operators both in the United Kingdom and abroad. An example of one of these operators that was very typical of prewar operation was Wrightways of Croydon.

The company was formed in 1935 by Messrs Wrightson and Pierce. Wrightways took delivery of its first DH89, G-AEML, on 26 September 1936; the fleet later expanded to include G-AFEZ and G-ADAL, the latter an ex-Hillman Airways machine. Wrightways Air Service, as it was known to the public, operated scheduled services to Paris, Dinard and Le Touquet. It also operated a special charter service for both passengers and freight, an air ambulance, and what was known as the 'Flying Showroom', so called because Wrightways equipped one of its aircraft to show off customers' goods, flying them anywhere in the United Kingdom and overseas. The brochure stated that 'Such tours always receive publicity from the local press and arouse great interest. With Sales campaigns becoming ever more expensive, this is an invaluable service'. Remember that this was 1938!

In 1939 Wrightways was awarded the Cumberbatch Trophy for 'Reliability' by the Guild of Air Pilots & Air Navigators of the British Empire, presented annually for 'meritorious work in the past year' for efforts put into the service between London and Paris. The service on this route was started early on in the life of the company as an early morning flight between Croydon and Paris. Originally carrying just newspapers and freight, it was eventually changed to accommodate passengers in 1936.

The summer of 1939 saw many of the Rapide flights by Wrightways being taken up by the Army for co-operations duties. With the outbreak of war all the aircraft were moved from Croydon and 'conscripted' under the control of the Royal Air Force. In fact, with Great Britain's declaration of war with Germany on 3 September 1939, all civilian aircraft came under the control of the

*Below:*
**G-AFEP (c/n 6406) was taken from North Eastern Railways Ltd in March 1940 and impressed into the RAF, becoming X9387.**
*RAF Museum*

*Above:*
**A33-3 was one of seven DH89s impressed from airlines into the Royal Australian Air Force, equipping No 1 Air Observers School. The aircraft is carrying the radio call-sign 'VHCRP' on its side.** *Bob Cavanagh Collection*

National Air Communications organisations. All air services were stopped, but were gradually resumed on internal routes that were considered necessary for the war effort.

No 24 Squadron RAF received many of the civil Rapides between 8 September and the critical days of May/June 1940; some were used as courier aircraft between Britain and the British Expeditionary Force in France. Ten were lost during the evacuation of France, G-ACYM and G-ADNI written off in accidents, whilst G-AEAM, G-AEBW, G-AEPF, G-AEXP, G-AFAH and G-AFEO were abandoned during the withdrawal. G-ADNH and G-AFSO were destroyed by enemy action.

Most of the Rapides impressed for war service kept their civil registrations for some while, these eventually being replaced with military serials. In all 43 were impressed, and nine were used in India. In addition, some of the DH89s operated by the oil companies in the Near East are thought to have been used locally by the armed forces for communications work.

The National Air Communications organisation came to an end on 9 June 1940, replaced by a new organisation formed under the Chairmanship of Sir Harold Hartley on 27 June, the Associated Airways Joint Committee (AAJC). All the surviving airlines joined the AAJC, with the exception of Allied Airways (Gandar Dower) Ltd which had a similar arrangement already in force with the government. Allied Airways operated with two DH84s and two DH89s, G-ACZF and G-ADAH, and in November 1941 it received the last production airframe from Hatfield, G-AGDM. In 1942 G-ACZE was also allocated to Allied Airways to replace G-ADAH which was under contract to the government. All Rapides used on scheduled routes were camouflaged, with the main cabin windows blacked out.

Rapide production had been stepped up since the outbreak of war, and the requirement for all of Hatfield's factory space to be used for full-scale production of the DH98 Mosquito meant that construction of the DH89 was transferred to Brush Coachworks Ltd at Loughborough.

From January 1941 the RAF's 'military' Rapide was renamed DH89A Dominie, the type serving in two forms. The wireless and navigation trainer was to be the Mk I, with the 8/9 seat communications and transport version the Dominie Mk II. In both these roles the aircraft was unarmed. The Dominie I could be easily identified by the roof-mounted direction finding aerial, and played a leading role in the training of navigators and radio operators for the RAF and Commonwealth Air Forces.

One unit that made particular use of the Dominie trainer was No 2 Electrical & Wireless School. This unit received its first Rapides during 1939, the early aircraft being equipped with R1082 receivers and T1083 transmitters for service with the Communications and Radar Flight. This equipment used different sets of coils to enable the operator to 'tune in' to the various frequencies. Each receiver used about 32 pre-set coils, the transmitter considerably less, usually about four. When flying training took place in bumpy weather, changing of the coils meant many difficulties for the operator, who could spend most of his time either chasing errant coils around the aircraft, or in trying to plug one into the set hitting the ceiling instead! Another problem was, according to some reports, that the equipment was 'useless when flying in close formation, but would happily raise Gibraltar "loud and clear" '.

Most of the pilots in the early days were either civilians or Polish nationals. During the latter stages of the war constant changes of aircraft meant that they were equipped with much better radio gear, the Marconi T1154 and R1155. This new equipment gave the ability to pre-select the frequencies in use, and coupled with the direction-finding loop wired to visual instruments, allowed the pilot to home in to a selected radio beacon.

The Commanding Officer of No 2 E&WS from June 1944 was Eric Farrington Birchall. His duties combined that of Air Instructor and control of the communications cabin staffed by RAF and WAAF personnel. Many pilots of this time were regarded as too old for front-line service, and the School even had a Church of England vicar as a pilot.

Each aircraft took five trainees, a typical flight lasting between one and two hours, and took the form of navigation exercises, either local or cross-country. Each student had to demonstrate the ability to communicate with home base and other stations up and down the country, obtain bearings and fixes, use the direction finding aerials and indicators, be aware of the emergency procedures, and at the same time keep a log of it all.

The first flight of the day would be at around 09.00hrs, the last at about 16.00hrs. Flying would last all day, and it was quite normal for the instructors to fly at least four details a day, involving five or six hours in the air. Sometimes an exercise detail would be combined with a communications flight to another aerodrome, which led to an interesting and varied flying programme. This all gave the pupils experience of reeling in and out the 250ft trailing aerial. The penalty for loosing the aerial, which could well happen on the approach if the student forgot about it, was 2s 6d (12½p). This was for the 24 lead weights on the end — in the wisdom of the Air Ministry the wire came free of charge!

Serviceability of the DH89s with No 2 E&WS was regarded as high, with very little problem even with the high usage. The Flight had 16 machines on strength at any one time, split into two flights of eight.

The DH89 was also used by other armed forces. Many flew in India and Africa, with aircraft assigned to the Kenya Air Auxiliary Unit and the Rhodesian and South African Air Forces. Rhodesia & Nyasaland Airways purchased VP-YAU in 1935, the first of five aircraft which were later taken over by the Rhodesian Government and used to form a combined airline/communications squadron until around March 1942. Known as the Southern Rhodesian Air Service, all aircraft carried military serials. In India the DH89s came under control of Air Command South East Asia, and were used for empire pilot and aircrew training.

*Above:*
**Operated by the Air Command South-East Asia in 1945, NR694 (c/n 6793) was Brush-built for the RAF.**

*Below:*
**Dominie X7382, probably in 1942.** *RAF Museum*

Wearne Air Services purchased Rapides VR-SAV and VR-SAW in 1937, which were named *Governor Raffles* and *Governor Fullerton* respectively. Both aircraft gave good service until Japan entered the war, when they entered the Malayan Volunteer Air Force, making up part of 'B' Flight. Both came to spectacular ends, VR-SAV destroyed in an air raid on Ipoh, VR-SAW struck by a taxying Hurricane at Palembang, Sumatra.

With the outbreak of war in Europe, the RAAF began a major training programme to provide aircrews. The twin-engined DH84 and DH89 were obvious choices for Navigation training and the RAAF ordered 87 DH84s because the Gipsy Major engines were already made in Australia under licence by General Motors Holden for the Australian Tiger Moth production line. The RAAF impressed seven DH89s from airlines to equip No 1 Air Observers School formed at Cootamundra, NSW, on 5 April 1940. These aircraft were given serials A33-1 to A33-7. All but one machine, A33-2, survived to be later returned to airline service.

In the United Kingdom many were assigned to Squadrons as communications machines, or assigned to the Metropolitan Communications Squadrons. The Air Transport Auxiliary also made much use of the type to assist the movement of ferry pilots in the delivery of other aircraft around the country for the RAF. Many of these DH89s were assigned to Headquarters ATA or No 1 Ferry Pool, both at White Waltham airfield at Maidenhead, Berkshire.

The RAF was not the only service organisation in Great Britain to use the DH89. The Royal Navy is reputed to have taken delivery of 64 aircraft, but since records for RN/FAA aircraft were destroyed in 1956, it has proved impossible to trace with any degree of certainty the service careers of these aircraft. Many of the Royal Navy Dominies were issued to 781 Squadron at Royal Naval Air Station Yeovilton and 782 Squadron at RNAS Donibristle.

One person who flew with 782 Squadron during the war years was Sgt Plt/1st Officer ATA Bill Byard. Bill, and Flg Off Hamilton, were loaned to 782 Squadron in 1943, carrying passengers and freight to naval air stations in Northern Ireland, Hatston in the Orkneys, Sumburgh in the Shetlands, and other small bases in support for the Navy. One aircraft Bill flew regularly was X7496, known as the Admiral's Barge. This Dominie was kept in top shape, with maroon leather upholstery and head rests. Bill remembers the start of one trip in X7496 with much amusement.

'I had several brass on board, and it was customary to pipe them on board. The Bosun did his duty, and I started the engines. An old Captain sitting in the seat beside the Radio Operator leaned forward and said "Good God, what a pair of lungs that Bosun's got!". Of course it wasn't the Bosun's pipe he was listening to, but the gyros running up!'

*Above:*
**One of the Dominies flown by Julian Courtenay Hunt, as described in the text. It is thought that this photo of Z7523 was taken at Burtonwood.** *Smithsonian Institute*

*Left:*
**Pilot Julian Courtenay Hunt in Dominie X7403 serving with the USAAF's 27th ATG.**
*Julian Courtenay Hunt*

'We were supposed to stay at Doni-B (as the station was nicknamed) for just July and August, but were requested to stay until at least October. The CO asked us to stay permanently offering me a full Lieutenant's rating.

'When we first arrived in June 1943 we had just a few familiarisation flights, not being allowed to carry passengers until we had been through the Naval Instrument Advance Flying School at Hinstock in Shropshire. I still don't know why, as we had no aircraft that were fitted with beam systems. Our radios still worked on the old civil frequencies!

'Back at Doni-B we were soon hard at work. On 27 June 1943 we flew Dominie X7496 from Doni-B to Eglington-Belfast-Prestwick, returning to Doni-B, then took Martlet IV FN296 to Renfrew on the 29th, then had lots of Dominie flying.'

It was not all easy flying, as the weather could turn bad very quickly. Bill's worst flight is described in his log as follows. It took place on 23 September 1943, and the aircraft was X7497 with five passengers for Hatston in the Orkneys.

'Worst flight ever. Told to approach Hatston east of Scapa Flow, permissible in bad weather only. Landed at Hatston 4/10 cloud on deck. Warned of showers at Sumburgh. Could not see through windshield. Result one flooded magneto. Stay overnight to dry out.'

The weather was also the cause of another interesting flight, this time on 20 September.

'Dominie X7486 Radio Op W/O Baxter and five passengers. Doni-B-Hatston-Stornoway. Force 4-5 gales all day. From NW. Course to Stornoway 270 was holding 315, dodging thunderstorms all day.'

Some of Bills flights played significant roles in wartime events: 23 September 1943 he and Leading Airman Amos flew Admiral Huntsvedt of the United States Navy from Doni-B to Hatston where the Admiral boarded the aircraft carrier *Wasp* off Scapa Flow to command a US/British task force to attack German shipping off Norway. This was the first time that an American had commanded major units of the Royal Navy.

Other important events came into Bill's flying career. On 11 October 1943 he picked up a Naval Lieutenant from Department M5 (Naval Intelligence) in Renfrew and was told to get him to Hendon so that he could be at Admiralty House by midnight.

'The Radio Operator called Speke (Liverpool) asking for refuelling, which was granted. Upon landing we were told that the 87 octane fuel was over the other side of the aerodrome. After an hour's delay we were off on our way again. Then my radio op informed me that Hendon was closing down due to an air raid warning and fog. I elected to go to Staverton (Gloucester) so that my passenger could get the 19.15hrs express train to London, and informed Hendon of my choice.

'We landed at Staverton in thick mist to be told by the duty officer that before we had arrived he was called by Hendon that an unidentified aircraft was heading their way, and to get the balloons up. Although we hadn't seen them, we had landed through the balloon barrage!

'The CO of Staverton came over and I explained that my passenger had to catch the train as he had important papers to get to London, but the CO told us not to worry and invited us both to dinner in the Mess. At the end of the meal the Lieutenant requested the use of the telephone to call Admiralty House. I went out with him, and he talked for a while before asking me to get the CO, who came out of the mess and was handed the phone. What a difference in attitude prevailed. "Yes Sir, Certainly Sir. At Once". I asked the Lieutenant who was on the other end — "An Admiral" was the answer.

'Within 20 minutes a car was waiting at the door, with orders to drive him direct to Admiralty House.

'A week later the CO of Donibristle called for me, showing me a commendation saying that I had been mentioned in dispatches, thanking me for getting my passenger to London after all the difficulties we had.

'I later found out that the Lieutenant was supposed to be carrying papers relating to a midget submarine raid on the German Heavy Water Plant in a Norwegian fiord.'

Throughout the entire period of flying

at Doni-B many other important people were carried, including Brigadier General Viscount Bridgeman from Turnhouse to Belfast, Surgeon Captain Breaks from Belfast to Machrihanish and United States Colonel Barr from Doni-B to Renfrew. Bill Byard's secondment to the Fleet Air Arm finished on 30 October 1943, with only one flight cancelled through mechanical problems when a radio set went US.

With the formation of the United States Army Air Force Eighth Air Force in early 1942 and its subsequent arrival in Great Britain during April, it was decided by Headquarters Command that there was a requirement for a light transport aircraft. This was met by the six DH89s loaned from the RAF and assigned to the Eighth Air Force Headquarters Command.

The first to be loaned were X7043, X7533 and X7523, all transferred to the USAAF on 1 December 1942. Three additional aircraft were loaned later, including V4724, ex-G-AFNC, a prewar Rapide. All were allocated to the Eighth Air Force Ferry & Transport Service, later the 27th Air Transport Group, a detachment of which operated DH89s from the base air depot (BAD) at Warton near Blackpool.

Flg Off Julian Courtenay Hunt was one of three Dominie pilots based with BAD 'A' at Warton, the others being Jim Martin and Tommy Duke, who flew the Dominies as freight and personnel transports. One very important aspect of these duties was to act as 'air taxi' for ferry pilots going to and from East Anglia, where most of the American bases were located. Julian takes up the story:

'I remember one of the pilots always carried a cribbage board and sat playing cribbage on the floor all the way back to the depot. There were no seats in the main cabin, so while I flew the corridor reaching from Kettering to Warton he played.'

The impression gained from this is that the work was very boring, if necessary. There *were* interesting moments.

'Jim Martin, Tommy Duke and I were on our way back to Warton in three Dominies. Duke and I were about 5min behind Jim. Tommy spotted Jim down

*Above:*
**X7454 *Wee Wullie*, the only USAAF Dominie to wear 'nose art'.**
*20th Fighter Group Association*

*Below left:*
**The RAF's RL959 at RAF Dishforth in October 1947.** *Eric Spurrier*

*Below:*
**The invasion stripes under the fuselage date this scene of an Ambulance Dominie (with Mrs A. Duncan ATA the unfortunate passenger).** *Mrs A. Duncan*

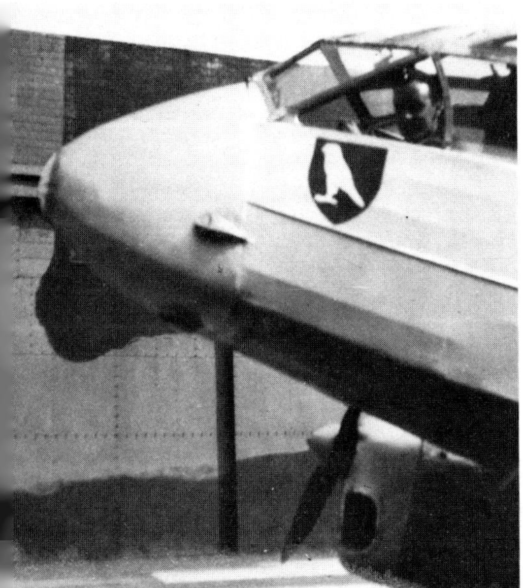

in a small field near a gun emplacement, Tommy circled, looking over the area before putting down his flaps and landing. Not to be outdone I landed also. Both Tommy and I landed in a field alongside the one Jim landed in, a slightly better choice as it turned out.

'It was an engine problem that forced Jim Martin down, so Tommy, myself and a guncrewman helped Jim and his crew chief roll the aircraft over a wooden bridge crossing a ditch into our field. After tying down X7523 for the night Tommy took Jim, and I took his crew chief. We backed the aircraft into a corner of the field, held the brakes, opened the throttles and blasted out of there. We circled the gun crew, rocking our wings in thanks, and headed of towards Warton.'

All of the DH89s serving with the USAAF wore the standard Olive Drab colour scheme, with light grey undersides. US markings were applied in the normal places, with the serials in black. No records exist to show if any US serials were ever allocated to these machines, only RAF serials being used. X7454 is known to have worn its serial in either white or very light grey, with the name *Wee Wullie* on the nose.

Two of the Rapides impressed into the RAF and used by Airwork Ltd as navigation trainers were later converted to ambulance aircraft. These were G-AFMH and G-AFMJ and had been subscribed for by the Silver Thimble Fund and named *Women of Britain* and *Women of the Empire* respectively. Both aircraft were handed over to the RAF by Her Highness Lady Maud Carnegie at a ceremony at RAF Hendon on 21 May 1941, and both aircraft were used in Scotland, *Women of Britain* surviving the war to be restored to the register as G-AFMJ in 1947.

The type was also shipped to New Zealand for service with the Royal New Zealand Air Force, which operated a detachment of DH89s in Fiji, possibly the most exotic setting for any use of the DH89.

1944 saw a communications flight set up by the RAF for the Dutch air force. The aircraft used were later transferred to the Netherlands after the liberation, soon to be joined by four other Dominies converted to civil standards for KLM, which was so keen to start up services that even as the task of liberation was continuing, a building was repainted at the ruined Schipol Airport near Amsterdam, domestic services starting in September 1945.

*Above:*
**The entire KLM fleet, in August/September 1945.** *KLM*

*Below:*
**PH-RAA is believed to have operated the first KLM service from KLM/Schipol on 6 September 1945.** *KLM*

# The Rapide Postwar

With the war in Europe almost over, military order for all types of aircraft were drastically reduced. This was indeed the case with the Dominie. Orders for the type were altered as early as November 1943, with a reduction of 60 airframes from a batch of 100. A later order for 80 aircraft was abandoned after the 20th built. The last 100 or so constructed at Brush Coachworks did not in fact see military service, although they were issued RAF serials. Along with many other surplus RAF aircraft they were sold off by auction at Nos 5, 7 and 18 Maintenance Units. A Dominie to Rapide conversion in many cases took the form of a new coat of paint, cabin soundproofing and the fitting of civil-type seats.

Depending on the civil application, the surplus Dominies were converted to one of two standards. A pilot-only operation, with seating for eight passengers, became the DH89A Mk 3, or the more unusual pilot and radio operator/navigator with six passengers was the DH89A Mk II. Although the

*Above and below:*
**These two views show G-AHKV in Automobile Association service. This aircraft was the 'prototype' for the DH89A Mk 6 conversion incorporating extensive and sophisticated electrical and navigation equipment.**
*Automobile Association*

attempt to perpetuate the mark number system — for both original and with later modifications — was made, it was seldom used.

Meanwhile the run down of the RAF gave surplus capacity, and many Dominies were put to good use giving air cadets their first taste of flying on 'Air Experience' flights. Flt Lt C. S. Walker, based at RAF Halton during this period, recalls:

'The Dominies from Halton ranged far and wide over the South of England to anywhere there was an airstrip we could get in. A day's work on Air Experience flying meant getting 80 air cadet bodies aviating. Depending on the weather the length of flight could, and did, vary considerably. It could be anything from a quick circuit of the field to a flight of up to half an hour. The main thing was to fly 80 bodies a day.

'The Halton Dominies could be away from base for weeks at a time, depending on the flying required. Pilots tended to shy away from the Dominie, not because of any difficulties, but because it was regarded by many as too easy to fly, and therefore did not stretch a pilot's ability.'

Soon civil aviation began to stretch its wings in Great Britain again. With Dominie conversion in full swing, many civil aviation concerns were snapping up these war surplus bargains. The Lancashire Aircraft Corporation, W. A. Rollasons, Brooklands Aviation, Field Aircraft Services, Air Schools Ltd and many others all brought Dominie conversions.

The British European Airways Corporation (BEAC) came into being on 1 January 1946, and was officially established on 1 August. It was Government policy to squash all independent airlines so that a state monopoly could be formed, both on internal routes as well as those in Europe. Two airlines fought a bitter struggle against nationalisation, but Channel Island Airways gave up the fight on 1 April, and Allied Airways (Gandar Dower) Ltd went to the wall on 12 April.

It was not long before the Rapides began to appear in the silver, red and grey scheme of BEAC, but it was soon realised that the company had far too many aircraft for its needs, so surplus ones were sold off via Airwork Ltd.

Nineteen of the remainder were brought together in 1950 and given the collective identity of 'Islander' class, being put to very good use on the Scottish, Scilly and Channel Island routes, many aircraft being given individual names.

Almost every postwar DH89 was powered by surplus RAF Gipsy Queen III engines, although one machine was converted by Flightways of Eastleigh in 1953 to take the Queen II powerplant driving constant-speed propellers. This prototype installation was made to G-AHGF, being promptly sold to New Zealand as ZK-BFK. With the Mk II engines the DH89 was cleared to take-off with an increased AUW of 6,000lb with a positive increase in climb, cruise and single-engine performance. Many other machines were later modified to this standard, easily identified by enlarged air intakes, variable pitch spinners and a more noisy take-off roar.

A one-off conversion was made by de Havilland to one of its own aircraft, G-AHKA, in 1948. This was fitted with manually-operated variable pitch propellers and special Gipsy Queen III MRP engines. It was retrospectively known as the DH89A Mk 4.

The last real attempt to modernise the DH89 came in 1956 when efforts were made to overcome the growing restrictions placed on commercial operators when flying in airport control zones. To achieve this, Air Couriers Ltd of Croydon converted G-AHKV with a sophisticated array of radio and navigation equipment, consisting of a Bendix SCR269A automatic radio compass, Murphy and Ekco VHF sets and three other pieces of radio gear, including a fan marker receiver. All this

*Above:*
**Brush-built for the RAF, c/n 6843 was registered G-AKZV for BOAC postwar. Later in its career it was sold to ZONDA for Argentina as LV-AGW, but its subsequent service and fate are unknown.** *Author's Collection*

*Right:*
**Wolverhampton Aviation's G-AKME was badly damaged by fire at Lymme in 1950.**
*R. Pairie*

extra electrical equipment put a very excessive demand on the aircraft's electrical system, far more than the wind driven generators normally carried could provide. To solve this the port Queen III was modified to drive a 1,000W generator producing 24V. G-AHKV was then sold to Ind Coope & Allsopp Ltd.

All this electrical power made G-AHKV the perfect choice for yet another 'first' for the DH89. It was later sold to Sky Neon Ltd, to be the only known example of a Rapide used for aerial advertising by means of fitting a framework under the lower mainplanes to which was attached a sign comprising a series of neon tubes.

Other Rapides were later converted with the same electrical fit as G-AHKV, along with the radio and navigational equipment. This conversion became known belatedly as the DH89A Mk 6 with the modification of G-APBM by Air Couriers Ltd in 1958.

Also in 1958, Fairey Aviation Co Ltd developed a special fixed-pitch metal

## Original Drawings of Cabin Options

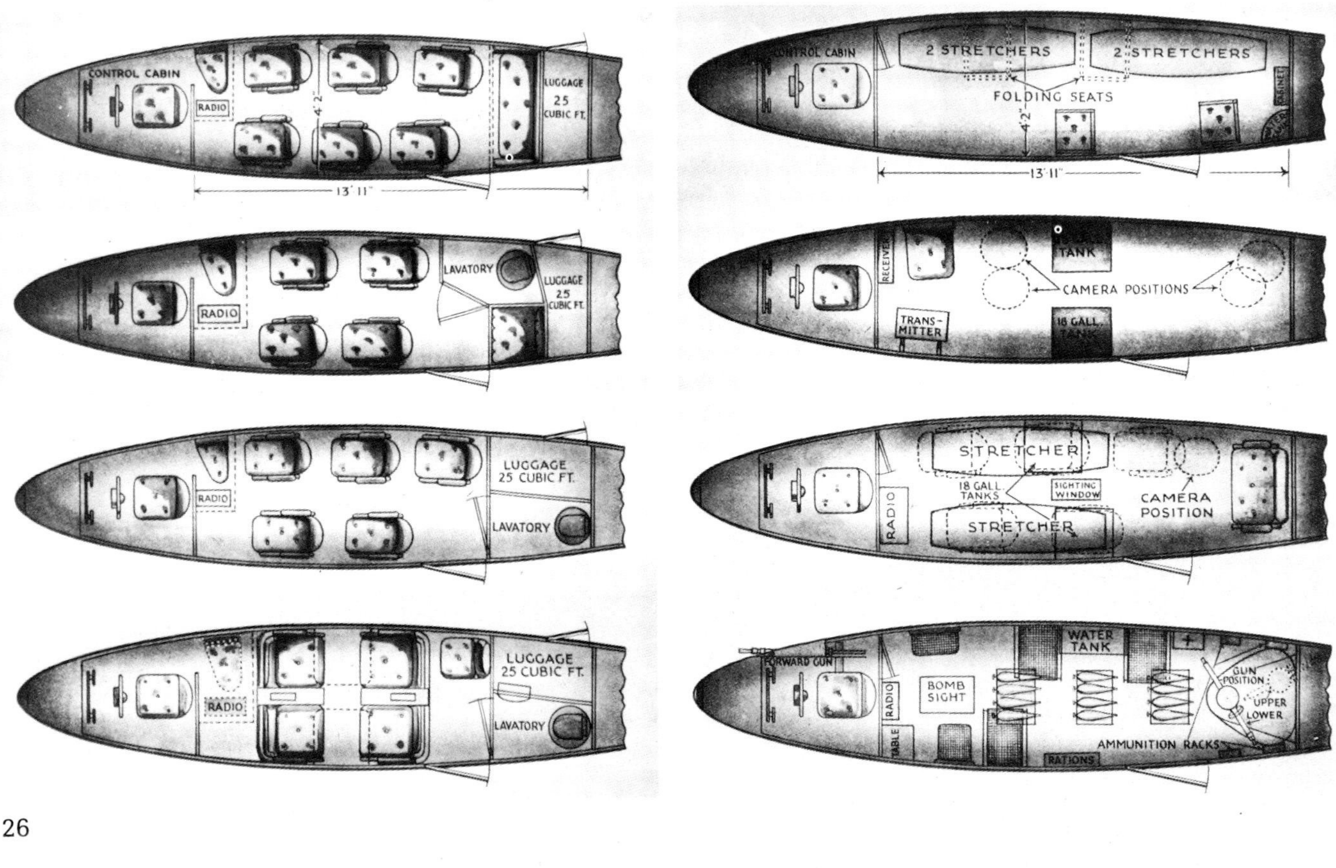

SPIRIT
20
HYK
308

*Above:*
**This photograph is a rare view of Belgian Air Force Dominie D-2, ex-NF874.**
*Musée d'Air, Brussels*

propeller for the DH89. Designated 'X-5', it imparted performance similar to that of the Mk 4 conversion without the need for the increased cost and maintenance complexity of the variable-pitch mechanism. Fairey itself used three DH89s based at the old ATA Headquarters at White Waltham — Fairey's own airfield had been taken over to build the new airport at Heathrow. The Rapides were used to carry ground staff and pilots around to the airfields of Boscombe Down, Bedford (Thurleigh) and Farnborough in support of the Gannet, Fairey Delta and Rotodyne projects. One of Fairey's pilots, Miss Lettice Curtis, once of the ATA and later author of the excellent book *The Forgotten Pilots*, remembers that apart from flying around England, the aircraft were used to commute to and from the associate company at Gosselies in Belgium.

Fairey's G-AHXW was fitted with variable-pitch propellers, which gave remarkable take-off characteristics. Lettice recalls that the aircraft was straightforward to fly, the only incidents being when pilots applied the brakes too hard and stood the aircraft on its nose. 'But I'm glad to say, that never happened to me.'

Many of the aircraft attached to Air Command South East Asia found their way on to the Indian civil register, serving with Indian Air Survey & Transport Co Ltd, Air Services of India Ltd and Vardraja Airways Ltd. Also in the Far East, Malayan Airways acquired two of the ex-BEAC machines in 1952 to operate third level services from Malaya to Borneo. A third aircraft was added to the fleet in 1955, and despite the rigours of both inhospitable terrain and tropical climate, all three survived to be taken over in 1958 by Borneo Airways Ltd.

Towards the end of the war negotiations took place between Southern Rhodesia, Northern Rhodesia and Nyasaland to form Central African Airways, which came into being on 1 June 1946. Some 12 DH89s were used, all ex-SRAS machines, the last being sold in 1950. Rapides were also used in the area by Zambesi Airways, Victoria Airways and Fishair Pty Ltd.

Eight aircraft were shipped to Peru and re-assembled in the Government factory at Callao, the first three climbing to a height of over 17,000ft to overfly the Andes to their new base in the hinterland. Two others were placed on de Havilland Canada floats, one being used at a naval base near Lima,

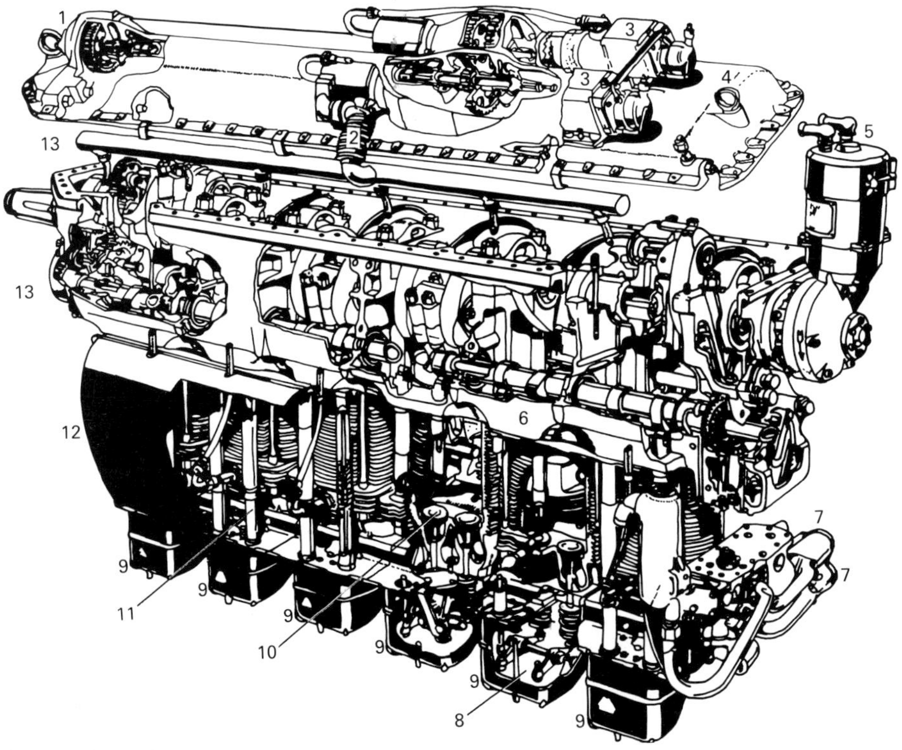


**De Havilland Gipsy Queen III Aero Engine Sectional View**
*Taken from lubrication diagram, DH Queen handbook*

**1** Dual Magneto Drive from Crankshaft — Single Drive from Crankshaft split at Magnetos; **2** H/T Leads (one set each side); **3** Magnetos; **4** Engine Lifting Eye-bolts (one each end); **5** 'Rotax' Electric Starter; **6** Camshaft; **7** Oil Pump and Filters; **8** Rockers; **9** Rocker Boxes (to be filled to level with oil); **10** Valves; **11** Push Rods in Tubes; **12** Cooling Air Ducting; **13** Camshaft Drive Crankshaft.

the second being used on the Amazon river.

British aircraft were also operated 'out of the country' for many years. Gibraltar Airways was re-formed in 1947 when two Rapides were leased from BEAC to provide links with Spain and Morocco. G-AKDX and G-AKDW arrived in mid-August, the inaugural flight to Tangier being made with G-AKDX under the command of Capt Arthur Brown on 30 August. Services to Tetuán, Seville, Malaga and Granada were later added using the company's own aircraft. The DH89 remained in service with Gibraltar Airways until 1953, when phasing out began in preference to the Dakota which offered greater passenger-carrying capability.

Morton Air Services operated several, including G-AGUG as far afield as Pakistan and India. G-AGUG was also leased to Standard Vacuum Oil Co on oil-related support work.

The Companhia de Transportes Aeros Ltd of Lisbon bought three in 1946. CS-ADI, CS-ADJ and CS-ADK received CoAs on 6 May, and were used on internal routes until early 1950, when all three were disposed of. CS-ADI was transferred to the Portuguese Air Force on 1 March 1950, when it was allocated the serial '2307', and operated as a communications machine until 1968, when it was retired to the Air Force Museum; because of shortage of space it is currently stored at Sintra Air Force Base. For many years this aircraft has been mistakenly identified as once being CS-AEB, ex-G-AGMA, but research in Portugal has proved this incorrect. The other two aircraft were transferred to Sao Tome e Principe on 27 February 1950 and re-registered as CR-SAD and CR-SAE. CS-AEB arrived in Portugal in 1950, but was transferred to Angola as CR-LCO on 28 March 1953.

The predecessor of Finnair, Aero O/Y, had DH89s OH-BLA and OH-BLB. Both aircraft were delivered before the war,

*Above and below:*
**G-AGUG on lease to the Standard Vacuum Oil Co Ltd for oil survey work in Pakistan and India.** *John Homewood*

*Top:*
**Sintra Air Force Base in Portugal at present stores '2307' (ex-CS-ADI) for the Portuguese Air Force Museum.** *Portuguese Air Force Museum*

*Above:*
**CS-ADK in service in Portugal with the Companhia de Transportes Aeros Ltd.** *Portuguese Civil Aviation Authority*

and somehow managed to survive until the late 1960s.

In Canada, airline amalgamation had been taking place, with 30 January 1942 seeing most of the surviving DH89s taken over by Canadian Pacific Airlines. Some independent airlines managed to survive, amongst which Queen Charlotte Airlines continued to operate DH89s, including CF-BND on floats.

Canada was also the origin of a 'mystery' DH89. In November 1941 fuselage shell No 89232 was built at Hatfield and shipped to de Havilland Canada along with other components as spares. This was later made up into a complete aircraft and issued the construction number CF-89232. A CoA was issued on 2 June 1942 and was registered CF-BNJ for Maritime Central Airways of Charlottestown until withdrawn from use and sold to Spartan Air Services in April 1952, its airworthiness certificate lapsing in 1955.

New Zealand had a number of DH89s on the civil register, many of them ex-RNZAF machines. West Coast Airways, Tourist Air Travel and Southern Scenic Airways all later became part of the new Mount Cook Airlines and all operated DH89s at some time or another.

Many South African Air Force machines were sold off locally after the war, prices varying from as little as £1,350 up to £3,913. The same thing happened in Australia when the Royal Australian Air Force-impressed machines were returned to civil use in 1942. In addition, on disposal, RAF surplus Dominies X7383 and RL947 were prepared for a 12,000-mile ferry trip to Alice Springs and service with Connelian Airways. Both were painted

Top:
**DH89 c/n 6401 OH-BLB *Lappi* was used in Finland as an air ambulance during World War 2, until the summer of 1941 when it was forced down by a Bf109.**
*Eino Ritaranta Collection*

Above:
**No 6401 seen at Eggemoen in 1964 after its withdrawal from use; it then carried the registration LN-BFB.** *Ole G. Nordbo*

in their Australian markings prior to delivery, X7383 becoming VH-BKM, and RL947 becoming VH-BKR. The ferry flight was uneventful until Darwin, when G-ABKM lost power on take-off, crashed into scrub off Daly Waters Airfield and burst into flames.

Life for the Rapides in Australia was far from easy. The hot climate and the proximity of fabric to engine exhausts made them particularly prone to engine start-up fires, an event that happened all over the world. VH-AAG fell foul of this problem on 12 December 1960 on a remote airstrip at Banyan Station in the New Territories. Flames from the starboard exhaust ignited the mainplane fabric, destroying both starboard mainplanes before the fire could be extinguished. To repair the damage, engineer Ivan Unwin was flown in from Queensland, spending the next three months camped in what was left of VH-AAG. Ivan worked right through the 'wet', and overcame a major setback to repairs. The replacement mainplanes were badly damaged when they were dropped into a swamp from under the helicopter whilst flying them in. The job was completed after 'eyeball' engineering to re-rig the aircraft for flight. VH-AAG was then flown to Darwin where Ivan completed more work to allow VH-AAG to pass its annual CoA.

Australian Rapides were used for all manner of tasks not attempted elsewhere. Darwin Air Taxis used VH-AAG for carrying buffalo meat until 1961, and two years later VH-AAG was written off in collision with a kangaroo, breaking the lower main spar in the process! Merv Ward, trading as Airflite, bought VH-BIF in November 1960, using it for dingo bait dropping and

grass seeding until 1963, when it was sold to the Northern School of Parachuting.

Possibly the longest journey ever made by a DH89 was the flight to Australia by VH-ECW. Built in 1941 as Dominie I X7370, it was eventually disposed of in 1947, going to BEAC as G-AJXB *Sir William Grace*. Sold to Sweden in 1956 as SE-CBU for Dragon Aviation AB, it was purchased by Petur Ahrens in 1960 as a cheap method of emigrating to Australia with his wife and family. Petur Ahrens was an experienced pilot and engineer, so no problems were expected. SE-CBU passed through Nicosia and Singapore, finally making landfall in 'Aussie' at Darwin after a trouble-free flight. The Ahrens family continued on to Sydney, before flying the aircraft down to Tasmania.

The last DH89 on the Australian register passed away in 1968, and for a time it was thought that the Rapide would never be seen in the skies 'down under' again. However, in 1974 two DH89s from New Zealand were imported, ZK-ALB becoming VH-IAN, whilst ZK-BCP was VH-BGP.

Back in Britain, although many of the smaller airlines had been nationalised as part of the British European Airways

*Above:*
**This picture of G-AGUF, the flagship of the Island Air Services fleet, was used as a publicity postcard by the company. The aircraft was destroyed in a crash at Ramsgate in June 1957.** *Monique Agazarian*

*Below left:*
**Island Air Services' owner and pilot — Monique Agazarian.** *Monique Agazarian*

*Below right:*
**G-AGUF, photographed from the company's G-AGJG, was being flown to RAF Benson by Monique Agazarian, for pleasure flights from the airfield.** *Monique Agazarian*

Corporation, the DH89 was still used by flying clubs, charter airlines and for pleasure flying. One company typical of this type of postwar operation was Island Air Services. IAS owned three Rapides, but leased others as the need arose. Owned by Monique Agazarian, IAS started operations in the Scillies, but commenced pleasure flying from Northolt when BEAC was based there just after the war. Later, operations moved to London Heathrow when a contract for pleasure flying from there was granted. The company's maintenance and administrative base was part of 'D' hangar at Croydon Airport.

*Above:*
**After wartime service as NR750, c/n 6838 was registered as G-AKIF. It was operated by Kentair Charters Ltd from 1966 to 1968, and was later used extensively for parachuting.** *Author's Collection*

*Below:*
**G-AEMH (c/n 6336) at Horsey Toll Airfield, Whittlesey in 1947/48 while with L. H. Hanson Air Charters Ltd. The aircraft is now preserved with Ian Jones of Babbacombe for restoration.** *Peter Steels*

Pleasure flights around London took two forms. A short circuit of the airport lasted 3½mins and cost 11 shillings (55p). The longer flight, around the Houses of Parliament, took 20mins and cost £1.

At least two of Island Air Services' aircraft were unique: G-AGJG and G-AGUF were the only two DH89s known to have been converted to carry nine passengers. This conversion was not without its setbacks: very little room was spare in the main cabin, and because of problems associated with the centre of gravity, great care had to be taken in loading. When flown with a full complement of passengers the fuel load was reduced to just 32 gallons, with no luggage allowed. Range was therefore somewhat limited.

G-AGUF had another claim to fame

while with IAS — it was stolen by an absconder from an Approved School. The aircraft was parked at Croydon Airport but the absconder managed to start one Gipsy Queen, careering around the airfield without enough power to take-off. Eventually the would-be pilot stopped the aircraft and made good his escape, going on to steal Auster Autocrat G-AGVL from Biggin Hill. Although not a pilot he managed to fly this aircraft across the Channel, making a crash landing in France. An ironic sequel to this was that Monique Agazarian flew the insurance assessors to France to view the wreck of the Auster — in G-AGUF!

Island Air Services also operated a daily scheduled service from Shoreham to Deauville with 100% regularity, and other charters were flown around England and Europe. Many flights were made to racecourses, including Aintree and Newmarket, ferrying jockeys from one race meeting to another.

Unfortunately, due to increasing traffic at Heathrow and the growing complaints against noise, something had to be done to appease the complainers, so in 1958 John Profumo, then Parliamentary Secretary to the Ministry of Civil Aviation, told the House of

Commons that all local flying would cease immediately. This rapidly brought about the demise of Island Air Services. It's hard to believe that the DH89 could be a noisy aircraft, especially when one considers that at this time the Boeing 707 was beginning to come into service at Heathrow. One suspects that it was not so much the noise as the difference in approach speeds that brought about this ban, and the larger airlines having more political clout than Island Air Services.

Although the Dominie had long since retired from service with the RAF, the 14 flying with the Royal Navy were still giving sterling service, mainly on air experience flights. This gave rise to a whole generation of ex-ATC cadets who will proudly state that their first taste of flying was gained in a Dominie at some long-past summer camp. These aircraft were still on charge as late as 1962, having rotated through storage at the Royal Naval Aircraft Repair Yard at Donibristle. They were sold off by public tender, four being 'civilianised', the remainder sold to Anglo-Diesel Co in June 1963, promptly disappearing without trace to defy all attempts at locating them.

British European Airways Corpora-

*Far left:*
**G-ALBI of the *London Evening News* at Fairoaks in 1956.**
*The late Alan Burfield/Author's Collection*

*Above:*
**G-AJBJ in a sorry state at Coventry, wearing the markings of Air Navigation & Trading.** *Stuart Howe*

*Below left and below:*
**The Berlin Communications Squadron was one of the many military operators of NR782, seen here with the Fleet Air Arm and in typical postwar FAA colours, with day-glo areas added on the nose and tail. The aircraft (c/n 6858) is carrying the Culdrose tail code.**

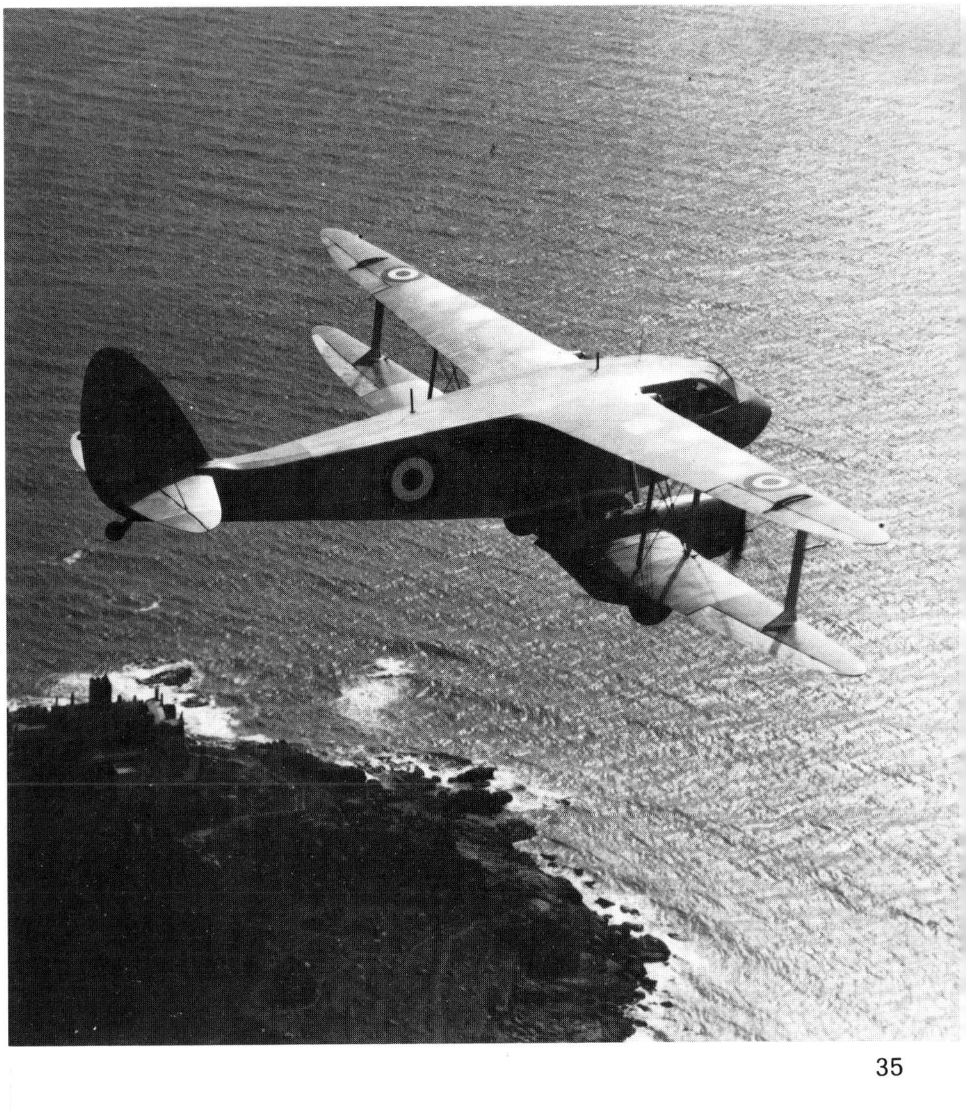

*Above:*
**Illustrating the type's position if the brakes were applied too hard is the RN's HG713 in standard postwar FAA colours.**

*Below:*
**Scillonia Airways was the last scheduled operator of the DH89 in the United Kingdom.**

tion operated its 'Islander' class Rapides on the Highland & Island routes for many years. The Scilly Isles route between St Mary's Airport on the Scillies and St Just on the mainland is a perfect example of the type of route that the Rapide was to call its own, and for many years the Rapide was the only type cleared for commercial operations into St Mary's. Indeed, the apparent ease with which the pilots flew in and out of the Scillies and some of the smaller islands of Scotland could have only been done with superb airmanship and the very forgiving nature of the aircraft.

During the summer months three machines were employed on the Scilly Isles route, reducing to one in winter. Some 20,000 passengers were carried annually, prefering the 20-minute air crossing to the three-hour sea journey by the steamship *Scillonian*.

During 1961 a new operator was authorised to run from Plymouth to Exeter then on to the Scillies. Mayflower Air Services started a scheduled service in 1962, initially with G-AHLM, but a second machine was added soon after the start of operations. Unfortunately G-AHLM swung on take-off from St Mary's and severely injured the owner of Mayflower, Capt Cliefe. G-AIUL, Mayflower's second aircraft, passed on to the Exeter-based British Westpoint Airlines, together with redundant BEAC machines G-AGSH, G-AHKV and G-AJCL. British Westpoint only operated for a couple of years before it too stopped flying.

BEAC kept the Scillies service operating with Rapides until 1964, when the last three were replaced with Sikorsky S-61N helicopters. This was not to be the end of Rapides in the Scillies, however, K. B. Neely bought three DH89s for his independent Scillonia Airways, and, operating the aircraft on routes around the West Country and the Scillies. Scillonia Airways had the distinction of being the last scheduled operator of the type in the United Kingdom.

# A New Lease of Life

Above:
**The free-fall parachutist's view of a Dragon Rapide.**

Although in the late 1950s the type must have seemed an anachronism to many passengers educated to expect jet aircraft, the Rapide soon found its own niche to live out the later years of its life. The big airlines may not have wanted it, and the many years of work with the armed forces was long gone, but the growing sport of sports parachuting found the type the almost perfect economical vehicle to lift bodies into the wild blue yonder.

In the early 1960s the Rothmans Sports Foundation, part of the Rothmans of Pall Mall tobacco concern, provided the Army Free Fall Parachute Training Association with G-AJHO, G-AGTM, G-AKIF and G-ASRJ. The first to be delivered was G-AJHO in June 1963. Purchased from Brooklands Aviation at Sywell where it had been used as a Directors' aircraft for many years, G-AJHO soon found its way over to Germany and the Rhine Army Parachute Association. She gained the nickname 'Groanin Grannie' but later acquired the more formal name of *Siegfried*.

G-AHAG was a latecomer to the sports parachuting field, when it was purchased by Monty de Cartier. This aircraft soon made its mark on parachuting in a big way — and not in the manner that some people would have liked — when it was the subject of an incredible accident on 15 July 1972, while based at Halfpenny Green. The aircraft formed one half of a pair of uplift aircraft for a 'big star' parachute attempt from 10,500ft in formation with a Britten-Norman Islander of Thurston Aviation. As the time for exiting the aircraft arrived, G-AHAG passed under the Islander. Mike Taylor, who had just exited the Islander, hit the fuselage of the Rapide and bounced off, breaking his femur in the process. Mike Bolton was not so lucky. He was next out of the Islander, and also hit the

Below:
**G-AGTM, photographed from G-AEML at Blackbushe in 1966, was one of the Rothmans Sports Foundation aircraft provided for the Army Free Fall Parachute Training Association.**

Rapide, but he crashed through the roof in the area of the emergency exit, smashing both his wrists. Mike managed to remain inside the aircraft until a safe landing could be made. This unique accident ably demonstrates the ruggedness of the design, and after repairs G-AHAG was soon back in use with Alex Black's parachute club at Lympne.

G-ALAX is one of the few Rapides to have been a film star. It was often used by Keith Tulloch and the Scottish Parachute Club, and made its film debut in Spain during 1966 when it was used in the parachuting sequences for Raquel Welch's *Fathom* made by 20th Century Fox. Raquel Welch played the part of Fathom Harville, a member of a team participating in the World Parachute championships. However, it was not all it seemed, as the Fathom who jumped from G-ALAX was really Ken Voss in a wig!

G-AGJG and G-AGTM were also used for parachuting along with G-AKIF, which was loaned to the Norwegian Paralift organisation from 1971-73 as LN-BEZ. D. W. Gray's ZK-AKU was flown in Rothman's colours in New Zealand, lifting parachutists 'down under'.

G-ADAH and G-AJGS were 're-discovered' in a hangar at Dyce, and in 1967 were taken by road to Booker Airfield where G-AJGS flew again in February 1968. Two years later it was ferried out to the United States from Halfpenny Green. G-ADAH, the ex-Allied Airways 'Pioneer', eventually went to the Royal Scottish Museum at East Fortune in 1974.

After spending a number of years in the USA, G-AJGS, one of a pair of Dominies that was completely rebuilt from spares just after the war by de Havilland, Witney, is back in the United Kingdom again, with Ron Souch at Hamble, undergoing extensive restoration.

The decline in use of the DH89 was halted in the early 1970s. Growing interest in vintage aviation by the general public ensured that the popularity of the Rapide was on the increase again, and made certain that those left would be well looked after. But several that really should have survived met with ignominious ends. G-AHKV, the ex-BEAC and Automobile Association aircraft, was burnt by the Birmingham Airport fire section despite efforts to save her in 1969, while VH-AAG and several others were burnt in Australia, again for fire practice. N8053, ex-G-ALGE/EI-AMN/F-BLXX, was restored in the early 1970s then exported to the USA. This aircraft is now only just in existence, having lain derelict with a broken back at St Maartin Airport in the Netherlands Antilles.

**The Preservation of G-AJHO**

With the halting of the type's decline in the early 1970s, the DH89 took on the new role of sedate, gentle flying with careful owners who were proud of what they had. It was now no longer considered simply as a workhorse, but looked upon as an aircraft that could be displayed in museums and airshows as an important part of the world's aeronautical heritage. Because of this, about 60 survived, just over half of which are displayed statically in aviation museums around the world, and the world's population of airworthy Rapides is slowly on the increase. The type is relatively easy to maintain, the major problem being the availability of powerplants. Luckily the Gipsy Queen III is relatively easy to obtain spares for, but — like everything else connected with old aeroplanes — at a price. In many cases the restoration of these machines is an uphill struggle, but every DH89 owner is at best enthusiastic, many almost fanatical!

One aircraft the author has been deeply connected with for a number of years is undergoing restoration at the present, and illustrates many of the problems that current owners and operators face.

C/n 6835 was one of hundreds built as Dominies by Brush Coachworks. It was allocated RAF serial NR747 and flown to de Havilland, Witney on 12 December 1944, then allocated to RAF Halton for Air Experience Flights and communications flying. NR747 was then transferred to No 18 MU on 10 May 1946 for conversion to civilian Rapide standard before sale by public tender.

*Above left:*
**G-AIYR of the Land's End Aero Club, photographed in the 1980s at the Famous Grouse Rally, Hatfield.** *Author's Collection*

*Top:*
**G-AJHO and G-AGTM fly low in formation over Netheravon.** *Author's Collection*

*Above:*
**G-AJHO with Brooklands Aviation Ltd.** *Barry Clarke*

It stayed there until 27 June 1947 when it was sold to Brooklands Aviation of Sywell, and registered as G-AJHO.

After spending many years with Brooklands — much spent shuttling between Little Staughton and Sywell for use by the directors, flying only as little as 20 hours a year — G-AJHO was sold to the Army Free Fall Parachute Training Association during June 1963. It was in the colours of the Rothmans Sports Foundation that G-AJHO was wheeled out for the first display with the Association, when a stick of eight free fallers was dropped at the Armed Forces day at Aldershot.

Having been purchased by the Association as an answer to the problem of providing an economic aircraft for the uplift of parachutists, G-AJHO showed a profit in the bank by the end of the first year. It had been borrowed by a pilot for a flight to Tangier, flown over 170 hours and dropped more than 1,000 parachutists at rates considerably lower than anywhere else in the country. It had not been all plain sailing, of course. G-AJHO had spent some time with the Rhine Army Para Association, and it was here that she was involved in an accident that ably demonstrates one of the few vices of the type. The crisis happened when one of the propellers was bent because the pilot had to make a sudden stop during taxying when some children ran in front of the aircraft. The tail started to rise, and both props touched the ground. Within 24 hours no less than five replacement propellers were on their way to Germany in response to urgent telexes. In the event none was needed, as a Germany-based de Havilland engineer put the blades in a vice and pulled everything straight in half an hour.

After a change of colour scheme, the aircraft was used in an all silver paint scheme, with blue registration letters.

In the summer of 1975 tragedy struck when the aircraft suffered a take-off accident at Netheravon when an attempt was made to get airborne with too full a load during hot weather. The aircraft suffered damage to the port lower mainplane, undercarriage and tail surfaces, and such was the low value of the Rapide at the time it was declared by the insurer as uneconomic to repair. The remains were placed at the rear of one of the hangars at Netheravon until they could be dealt with. The insurer then opened negotiations with East Anglian Aviation Society Ltd.

EAAS Ltd had already been involved in the restoration of a P-51 Mustang, and had arranged two Spitfire rebuilds. Satisfying as it may be to work on these aircraft, those who had worked could only stand by and watch them fly. With the Rapide it would be possible for the members to go flying. Norwich Union Insurance eventually donated the remains to the Society, and on 2 July 1976 members 'descended' on Netheravon to dismantle the remains for transportation by road to their base.

Upon full inspection of the airframe the damage was nowhere as bad as was first thought. Three of the four mainplanes were perfectly sound; the fourth, suspected by the insurance company as having a broken main spar, was in fact found only to be bruised when the wing structure was opened up. A repair for the damage was listed in the maintenance manual, along with repairs to the leading edge.

The years of parachuting had taken their toll on the fuselage; the seats had been removed years before, but have been located and replaced. The two *(continued on page 42)*

# The Restoration of G-AJHO

The East Anglian Aviation Society Ltd received a suggestion from Air Vice Marshal J. de M. Severne, Captain of the Queen's Flight, Royal Air Force, to consider placing its aircraft in the colours and markings of the first aircraft used by the then King's Flight in 1936 to commemorate the 50th anniversary of the Flight in July 1986. AVM Severne sought and obtained permission on behalf of the Society for this scheme to be used.

For a representation of this importance, not only must the correct colours be displayed, but the original registration must be used. Special dispensation was sought from and granted by the Civil Aviation Authority, for G-AJHO to fly carrying the registration G-ADDD for the celebrations. A scale model of the original aircraft was already on loan to the Queen's Flight from the Royal Air Force Museum, and this was used to determine the correct colours and marking positions. In particular, it proved to be of much use in determining the correct shades of red and blue.

In April 1986 the Queen's Flight took delivery of the first British Aerospace 146, and it had been suggested by AVM Severne that the Rapide should be

*Above:*
**G-AJHO in 1985, without the upper wings and lower extension planes.**
*All photos by the author*

*Below left:*
**The workers (from left to right): Vic, Rod, Anne (the author's wife) and Martin.**

*Below:*
**Repairing the damage to the port lower extension plane during January 1986 (and yes, it is cold!).**

*Above right:*
**February 1986, and three of the four mainplanes are fitted — G-AJHO is beginning to look like an aircraft once again.**

*Above far right:*
**A view from the cabin in February 1986.**

*Right:*
**The starboard stubplane ready for covering.**

*Below right:*
**The underside of the main cabin floor, showing the cables and controls.**

*Below far right:*
**Here the cabin is almost ready for the installation of the new instrument panel: the pilot's seat has been removed for ease of access.**

present. However, despite the efforts of all the people involved this was not possible.

To assist the Society in its endeavours, Ken Wheeler, Joe Cupit, Roger Blaikie and the apprentices in BAe's Modular Training Section at Hatfield offered much practical support.

An attempt is to be made to have the Rapide present at RAF Benson, home of the Queen's Flight, for the Families Day on 26 July, also at RAF Benson, just five days after the 50th anniversary on 21 July.

The Rapide's pilot for all of these events may well be Monique Agazarian, once of the ATA and one-time owner of Island Air Services, currently operating Air Training Services near Heathrow.

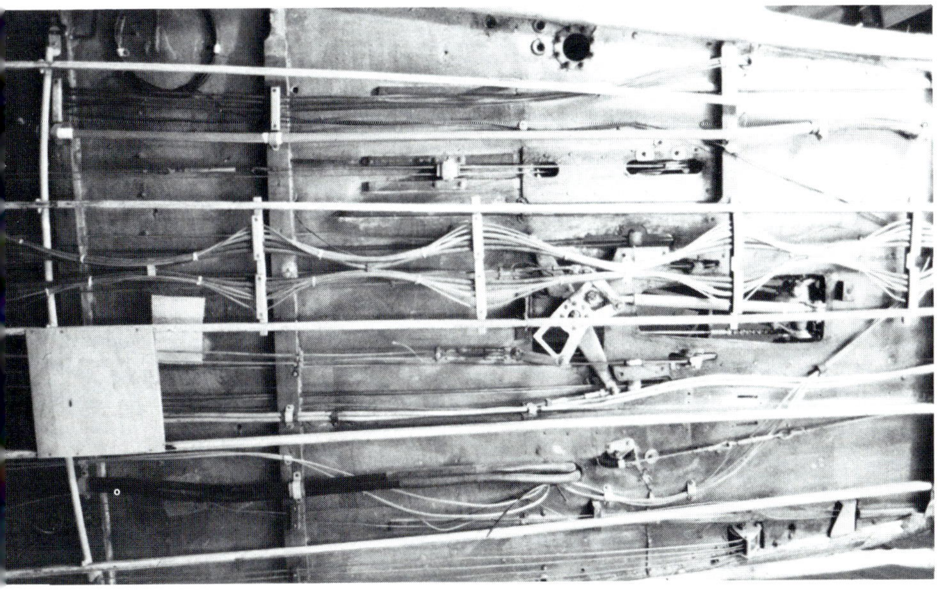

*Above:*
**G-AJHO dwarfed by Laker Airways' DC-10 at RAF Gutersloh.** *Peter Thorne*

*Right:*
**The attractive Rothmans of Pall Mall colours on G-AJHO.** *Author's Collection*

*Below:*
**Course members of the Army Free Fall Parachute Training Association push Rothmans-sponsored G-AJHO *Siegfried* (ex-NR747, ex-*Groanin Granny*) out for take-off circa 1969. The aircraft is now the only surviving DH89 Mk II anywhere in the world.**

stub-planes removed from the airframe for transit were dismantled and stripped to bare metal for inspection, in places necessitating the removal of 14 layers of paint. The metal structure was re-proofed, and repairs to the timber structure made where required. The trailing edge of the port stub was particularly bad, having suffered years of bodies jumping from the step, and in some places the trailing edge woodwork had completely broken up.

The removal of the stub-planes also afforded the opportunity to inspect the spar tube running under the fuselage. The old operating problems have appeared again with increased usage, and inspection ensures confident re-assembly, especially as this has been the subject of a recent manufacturer's notice on inner corrosion. Some of the airframe items had been removed before it was collected from Netheravon, so it has been the job of every member of the Society to scour the world locating missing items and spares.

An equally difficult problem was in refurbishing the Gipsy Queen engines. One engine was out of hours, and had to be replaced.

The East Anglian Aviation Society is made up entirely of enthusiasts, not qualified engineers. Therefore airframe and powerplant engineers have been engaged to do the work beyond the capabilities of Society members. All of this takes tremendous dedication and large amounts of money. Every member must do their bit in raising funds, both for the restoration project and to cover the cost of building up a small spares holding.

**In Preservation**

The oldest surviving DH89 is CF-PTK, the fourth production airframe, which is owned by J. R. Bowdery of Newfoundland. The current status of this machine is unknown, but it is thought to be under restoration.

G-ACYR was used to ferry General Franco to Spain for the start of the Spanish Civil War. Presented to Franco during the 1950s, it is in display in the Spanish Air Force Museum in Madrid.

An ex-Allied Airways (Gandar Dower) Ltd aircraft is G-ADAH, residing with the Royal Scottish Museum. It was found in the back of a hangar at Dyce, and is a low-hours machine.

G-AEMH and G-AKRN are two ex-East Anglian Flying Services aircraft currently under restoration with Ian Jones.

C/n 6345/EC-AKO is owned by Metamar S.A. of Madrid. It is believed that this machine is still flown regularly.

CF-BND is still just extant: although crashed in July 1949 on Digby Island, Canada, the remains are owned by the Western Canada Aviation Museum and a cover has been built over the site. It is thought that the aircraft is more or less complete, and it is hoped to get the remains off Digby Island and to the museum's facilities in Winnipeg when funds permit.

D-IGUN is owned by the Air Classic Collection and displayed at Dusseldorf Airport.

An ex-Fairey Aviation aircraft, OO-CNP is under restoration at the Musée de l'Air in Brussels. The machine was excessively modified in the past, and is being restored to static condition.

The South African Air Force Museum has the remains of two machines,

*Below:*
**OO-AFG (c/n 6458) of Air Affairs of Gosselies, Belgium, seen about 1963, was an ex-Fairy Aviation aircraft and was re-registered OO-CNP in 1964. The machine has been excessively modified, and is now being restored to static condition.**
*Author's Collection*

ZS-DJT and ZS-DLS, but Capt Dave Becker states that the total remains do not amount to a complete aircraft, and is searching for more.

G-AGJG, once of Island Air Services, is under extensive restoration by the Duxford Aviation Society to static standard. But is reported recently to have changed hands to Messrs Mike Russell and Dave Miller. It will be restored to flying condition. Also at Duxford is Mike Russell's G-AGTM. Mike and his band of helpers fly G-AGTM most weekends, giving passenger flights. Still with a commercial CoA, it is painted in Royal Navy/FAA colours.

Jean Salis of France has F-BZCA, and flew it in the Paris to United States Air Race. The machine is named *Blue Way*, and is airworthy.

Another airworthy Rapide is N2290B, an ex-Indian aircraft. Owned by Mike Kimbrell of Washington State, the aircraft is flown regularly alongside Mike's DH84 Dragon G-ADDI. The Rapide is of special interest because it is fitted with non-standard nose glazing.

John Pierce owns G-AKOE and G-AJBJ. G-AKOE is airworthy, but no

*Above:*
**Mike Russell's G-AGTM, in RN/FAA colours as NF875, seen from G-MOTH.** *Stuart Howe*

*Below:*
**Mike Kimbrell's *Mary Linda*, c/n 6588, was built for the RAF and registered initially as X7466. After service in India it was sold to an American owner and gained its registration N2290B. The non-standard nose-glazing is noteworthy.** *Mike Kimbrell*

*Below left:*
**G-AGJG, once of Island Air Services, under restoration at Duxford in 1984.**
*Author's Collection*

*Bottom:*
**ZK-AKY has the unusual joint Royal New Zealand Air Force (with which it served as NZ525)/civil colour scheme illustrated here.**

details are known of G-AJBJ's condition.

The New Zealand Museum of Transport & Technology's ZK-AKS is currently under restoration for static display. Additionally, VH-BGP is an ex-New Zealand aircraft, currently owned, flown and displayed at Drage Airworld, Woodonga, Australia; ZK-AKY is under restoration with T. Williams of Te Parae, New Zealand; and the third New Zealand machine is VH-AIN, owned and flown by Ken Orrman of Victoria, Australia. Finally, ZK-AKU suffered a ground loop recently, but is under restoration with D. W. Gray of Auckland.

Viv Bellamy owns G-AIYR, which was flown to South Africa recently on a commemorative flight. Unfortunately it suffered an undercarriage collapse after it returned, and is currently under restoration with British Aerospace.

The Musée de l'Air in Paris has F-BHCD hanging from the roof with parachutists jumping from it, depicting its days with the Parachute School.

Doyle W. Cotton has N89DH, and flies the aircraft regularly with an experimental certificate from the FAA. The aircraft is painted in RAF markings,

*Top:*
**V-3 (c/n 6740) its authentic markings, on display at the RNethAF museum at Soesterburg AFB.** *Stuart Howe*

*Above:*
**N683DH's postwar service included a period with BEAC as G-AHXW *Sir John Nicholson*. Now in the care of the Experimental Aircraft Association, it has been restored to true BEAC markings.**

*Below left:*
**Brush-built, aircraft c/n 6706 eventually found its way to France in 1954 where it was registered F-BHCD. It is now displayed at the Paris Musée de l'Air.**
*Robin Ementon, EASS*

*Below:*
**G-AIYR under restoration by British Aerospace apprentices.** *Stuart Howe*

*Above:*
**John Schell's attractively painted N2290F, an ex-Indian machine. Like N2290B it has the non-standard nose-glazing.** *Mike Kimbrell*

*Below:*
**Michael Astor's airworthy G-AHGD *Women of the Empire* at Old Warden in 1985.**
*Author's Collection*

*Right:*
**F-BCBD before restoration commenced at the Aviodome Museum, Amsterdam.**
*Harry van der Meer*

*Below right:*
**The unidentified DH89 in a shed at Durne Airport, Antwerp.** *Harry van der Meer*

with its registration carried in the manner of a RAF serial. The aircraft is currently up for sale.

G-ALXT returned to the UK from Ceylon, and is at present painted in the colours of Railway Air Services, which it never wore. Owned by the Science Museum, it is kept at the out-station at Wroughton, near Swindon.

An authentic Royal Netherlands Air Force aircraft, 'V-3' is displayed statically at Soesterburg AFB in the correct colours and markings.

N683DH is owned and displayed statically by the Experimental Aircraft Association at its museum at Oshkosh. It is thought that the aircraft is missing a set of cabin seats.

In the same state as Mike Kimbrell's N2290B is N2290F, owned by J. Schell. Another ex-Indian machine, the aircraft currently has a Chinese Dragon painted on the fuselage.

John English has ZS-JGV in Port Elizabeth, South Africa: it is thought to be airworthy.

Sir Adrian Swire owns G-AKIF, A. F. Ward owns G-AIUL, and Michael Astor owns G-AHGD. G-AKIF and G-AHGD are airworthy, but G-AIUL's condition is unknown.

G-AGSH was up for sale recently, until purchased by John Davy of

*Above:*
**Sold to the Israeli Air Force Museum in 1978, 002 is ex-RAF RL981, c/n 6952, and is airworthy.** *Stuart Howe*

*Left:*
**G-AHED (c/n 6944) is 'owned' by Hunting Aerosurveys Ltd, but is in store with the RAF Museum at Cardington awaiting restoration as RL962.** *Author's Collection*

*Below:*
**NR713 (c/n 6801) began its life with the RAF's No 18 Maintenance Unit in late 1944. Postwar it passed into civil hands as G-AJSL, finding itself with the Parachute Association from late 1966 to early 1968 as *Pegasus*. It was rebuilt at Biggin Hill in 1978 after tipping on to its nose at Sunderland airport, and was subsequently sold to an Australian owner. Currently it is in store at Moorabin, Victoria, Australia.**
*Ken Tilley Collection*

Specialist Flying Services of Carlisle, to use 'for pleasure'.

F-BCBD was sold to the Aviodome Museum in Amsterdam, and is, according to Harry van der Meer, Technical Curator, 'under restoration to fly'. Harry van der Meer has also been involved in the discovery of a mystery DH89 found in a building near Durne Airport, Antwerp. The history of the aircraft is obscure — it could well be ex-D-2 of the Belgian Air Force — and its future is uncertain because its owner apparently does not want to sell, restore or swap it.

N8530 is, as has already been mentioned, derelict at St Maartin in the Dutch Antilles. Several owners have tried to save the aircraft, but since the machine was arrested for drug smuggling its ownership is in dispute. It is quite possible that this aircraft has been lost, and the salvage of parts for other projects is impossible.

'002', ex-G-AKRS, was sold to the Israeli Air Force Museum and flies occasionally: G-AHAG is under slow restoration with Ralph Jones of Southern Sailplanes in Wiltshire; G-AHED is with the Royal Air Force Museum as RL962 (its old serial); and G-AIDL is operated by Capt Mike Hood on pleasure flights.

# DH89 Dragon Rapide Technical Specifications

DH89 Dragon Rapide mark numbers were only used with any significance before the outbreak of World War 2. After the war an attempt was made to perpetuate the mark number system, but this never met with any great success. However, the full range of variants are of interest, so are listed below.

The DH89A Mk II and 3 were postwar conversions of the military Dominies Mk I and II. As can be imagined, some of these aircraft were then further converted to Mk 4, 5 or 6 standards, which would have led to much confusion if the mark number system had been implemented.

| | |
|---|---|
| DH89 | Basic model from prototype |
| DH89A | Airframe fitted with flaps and landing light |
| DH89A Dominie Mk I | RAF navigation and radio trainer |
| DH89A Dominie Mk II | RAF communications aircraft |
| DH89A Mk II | Airframe with pilot and radio operator |
| DH89A Mk 3 | Airframe with pilot and eight passengers |
| DH89A Mk 4 | Airframe fitted with Gipsy Queen IIs and constant speed propellers |
| DH89A Mk 5 | Fitted with manually operated variable speed propellers |
| DH89A Mk 6 | Fitted with Fairey metal X5 propellers |

**DH89 Rapide Leading Particulars**

*Dimensions*

| | |
|---|---|
| **Span:** | 48ft 0in |
| **Length:** | 34ft 6in |
| **Height:** (tail down) | 10ft 3in |
| **Wheel track:** | 11ft 6in |

*Weights (lb)*

| | **Mk II** | **Mk 3** |
|---|---|---|
| **Tare Weight:** (standard without seats) | 3,555 | 3,520 |
| **Pilot:** | 170 | 170 |
| **Fuel:** (76gal at 7.25lb/gal) | 551 | 551 |
| **Oil:** (8½gal at 9lb/gal) | 76 | 76 |
| **Balance available:** | 1,198 | 1,233 |
| **Disposable load:** | 1,995 | 2,030 |
| **Max all-up weight:** | 5,550 | 5,550 |

*Speeds*

| | |
|---|---|
| **Take-off safety speed:** | 75mph IAS |
| **Best climb speed:** | 85-90mph IAS |
| **Economic cruise speed:** | 130mph IAS |
| **Landing speed:** (flaps down) | 75-80mph IAS |
| **Stall:** | 62mph IAS |
| **Max speed:** | 157mph IAS |

*Below:*
**Derelict N8053 had an eventful career. It was converted post-RAF service for BOAC for Iraqi Airways, worked with a number of charter companies and was sold to a French owner in 1964, as F-BLXX. After restoration in the UK it was painted in British European Airways colours as *Lord Shaftesbury* by its new US owner. Later it was impounded, having been used for drug smuggling, and was badly damaged when hit by Hurricane David in 1979.** *Author's Collection*